The Infinite Abyss Series:
Ignite Your Passion for Learning.

Stimulate Your Curiosity with Astounding Facts and in-depth Exploration of the Human Body, the Oceans, Outer Space, and the Cosmos.

House of Abundance Publications

Table of Contents

The Abyss Above

Chapter 1: What is Astrology? 1

Chapter 2: Galaxies - Vast Cosmic Islands 6

Chapter 3: Exploring the Universe - From Dwarfs to Supergiants..14

Chapter 4: Dark Matter and Cosmic Mysteries...........................20

Chapter 5: Astounding Phenomena in Space 25

Chapter 6: Unveiling the Local Group: A Galactic Ensemble of Marvels and Interactions 29

Chapter 7: Cosmology - The Study of the Universe.......................35

Chapter 8: Astronomy vs. Astrophysics 40

Chapter 9: Amateur Astronomy - Unveiling the Transient 46

Conclusion 51

Epilogue 54

Resources 57

The Abyss Below

Introduction 63

Chapter 1: An Oceanic Overview 65

Chapter 2: The Etymology of 'Ocean' 68

Chapter 3: Origin of Water and Formation of the Ocean............. 71

Chapter 4: The Ocean's Role in Planetary Formation.................... 77

Chapter 5: The Ocean in the Climate System............................. 81

Chapter 6: Geography of the Oceans 88

Chapter 7: Dive into the Ocean's Layers........................... 95

Chapter 8: Oceanic Temperatures and Currents.......................... 98

Chapter 9: Gas Exchange in the Ocean........................... 104

Chapter 10: The Ocean's Rich Biodiversity 109

Chapter 11: Legends of the Deep 114

Conclusion 118

Resources.. 121

The Abyss Inside

Introduction 127

Chapter 1: Human Anatomy 129

Chapter 2: Circulatory System 138

Chapter 3: Digestive System 150

Chapter 4: Respiratory System 158

Chapter 5: Nervous System 165

Chapter 6: Reproductive System 179

Chapter 7: The Integumentary System 189

Chapter 8: The Endocrine System 195

Chapter 9: The Immune System 201

Chapter 10: The Urinary System 208

Chapter 11: Mind-Blowing Facts and Discoveries 214

Conclusion 222

Resources 228

The Abyss Above:

Mind-Blowing Facts About Astronomy, the Cosmos, and Outer Space

Imagine gazing at a midnight sky, where the cosmos unfolds before your eyes like a celestial tapestry. Each twinkling star, every mysterious planet, and the vastness of space itself hold secrets waiting to be revealed. Welcome to "The Abyss Above: Mind-Blowing Facts About Astronomy, the Cosmos and Outer Space," an awe-inspiring journey through the wonders of astronomy.

Have you ever wondered what lies beyond our blue planet? Are you curious about the enigmatic forces that shape galaxies or the mind-boggling phenomena that occur millions of light-years away? If so, prepare to embark on an extraordinary quest for knowledge.

In this captivating book, we'll embark on an interstellar expedition, delving into the depths of space and time. We'll unveil the mind-bending facts that will make you question the very fabric of our universe. Every chapter will reveal astonishing revelations, from the birth of stars in colossal nebulae to the earth-shattering death of massive supernovae.

But this journey isn't just about scientific discoveries. It's about the stories woven into the cosmos. We'll reveal the ancient tales of constellations that have captivated humanity for centuries. We'll explore the myths and legends that bridge the gap between science and imagination, reminding us that the beauty of the cosmos is as much about wonder as it is about knowledge.

As we explore the grandeur of black holes, where gravity bends time and space. Witness the intricate dance of planetary systems, where the possibility of extraterrestrial life looms tantalizingly close. Marvel at the cosmic ballet of galaxies colliding, giving birth to new stars, and reshaping the very face of the universe.

"The Abyss Above" will ignite your passion for the cosmos through vivid storytelling and expert insights. Whether you're an aspiring astronomer or simply an admirer of the night sky, prepare to be captivated by the wonders that await.

So, fasten your cosmic seatbelt, for the journey into the infinite expanse of the universe is about to begin. Get ready to embark on an adventure that will expand your mind and leave you in awe of the vast, magnificent celestial mosaic we call the cosmos.

Chapter 1:
What is Astrology?

Definition and Exploration of Astronomy As a Natural Science

Astronomy, a natural science, involves the study of celestial objects and phenomena in the vast universe. Astronomers seek to understand the stars, planets, galaxies, and other cosmic entities through scientific methods and tools. This observational science utilizes telescopes, satellites, and advanced instruments to collect data of the cosmos.

Astronomy primarily aims to trace the origins and evolution of celestial bodies and the universe itself. By analyzing the light emitted or reflected by distant objects and observing changes in their properties and positions, astronomers clarify the cosmic timeline and gain insights into the universe's early stages.

Astronomy also contributes to understanding the physical laws governing the cosmos. By studying the behavior of astral bodies, astronomers test and refine theories related to gravity, electromagnetism, and nuclear physics. Collaborating with physics, chemistry, and planetary science experts, astronomers explore the connections between planetary systems and the underlying physical and chemical processes.

As we explore astronomy as a natural science, we embark on a journey that reveals the wonders of the cosmos. From uncovering the complexities of distant galaxies to studying the planetary dynamics within our solar system, astronomers push the boundaries of knowledge and expand our cosmic horizons. Through their efforts, we gain a deeper appreciation for the immense beauty, diversity, and grandeur of the universe we call home.

The Distinction Between Astronomy and Astrology

As we dive deeper into the captivating realm of astronomy, it is crucial to clarify a common misconception: the distinction between astronomy and astrology. While both fields share a historical connection, they have evolved into distinct disciplines with different aims and methodologies.

As we have learned, astronomy is a scientific endeavor focused on studying cosmic entities. It seeks to understand the laws and mechanisms that govern the universe through empirical observation, mathematical models, and rigorous experimentation. Astronomy explores the physical properties, behaviors, and interactions of celestial bodies, shedding light on the nature of the cosmos.

On the other hand, astrology is a belief system that posits a connection between astronomical bodies and human affairs. It claims that the positions and movements of celestial bodies, particularly the Sun, Moon, planets, and constellations, can influence and provide insights into human personalities, relationships, and even future events.

Astrology often involves horoscopes, zodiac signs, and interpretations based on supposed astrological influences.

While astronomy relies on the scientific method, evidence-based research, and objective analysis, astrology is rooted in subjective interpretations and personal beliefs. The methods employed in astrology lack scientific rigor and are not subjected to empirical testing. Therefore, astrology does not adhere to the rigorous standards of evidence and verification upheld by astronomy.

It is crucial to differentiate between these two disciplines to avoid confusion and maintain the integrity of scientific inquiry. Astronomy delves into the vast cosmic wonders to understand the universe's mechanics. At the same time, astrology offers a belief system centered on the perceived influence of celestial objects on human lives.

In the following chapters of this book, we will continue to explore the captivating realms of astronomy, unveiling incredible facts about the cosmos, its celestial inhabitants, and the remarkable forces that shape our universe. Together, let us journey further into the depths of knowledge and wonder as we embrace the scientific marvels of astronomy and leave behind the realms of astrology.

Historical Significance of Astronomy in Different Cultures

Throughout the ages, astronomy has held immense significance in various cultures worldwide. From the earliest civilizations to today, humanity's fascination with the night sky has shaped our understanding of the cosmos and influenced cultural beliefs, rituals, and scientific advancements.

For instance, the ancient Egyptians developed a sophisticated astronomy system tied to their religious practices. They observed the movements of celestial bodies, such as the Sun, Moon, and stars, to develop calendars and align their religious ceremonies with astronomical events. The pyramids were designed with celestial alignment, showcasing the deep intertwining of astronomy and Egyptian culture.

Babylonians, renowned for their astronomical achievements, meticulously recorded celestial observations on clay tablets. They established a calendar based on lunar phases and developed mathematical methods to predict celestial events. Their knowledge of celestial motions laid the foundation for later astronomical advancements in Mesopotamia and beyond.

Ancient Greek civilization birthed notable astronomers like Aristotle, Hipparchus, and Ptolemy, who sought to explain the movements of celestial bodies and develop cosmological models. Greek philosophers pondered the nature of the universe and its place in the grand scheme of existence, leaving an indelible mark on the history of astronomy.

In India, ancient astronomers made significant contributions to astronomy and mathematics. The concept of zero, essential to modern numerical systems, originated in Indian mathematical texts. Astronomical observations in India were essential for religious and agricultural purposes and the development of precise calendars.

The Chinese civilization cultivated a rich astronomical tradition, observing and recording celestial events for centuries. They carefully documented supernovae, comets, and solar eclipses, attributing cultural and political significance to these celestial features. Chinese astronomers also developed accurate astronomical instruments and mathematical models, contributing to understanding celestial motions.

Indigenous peoples of the Americas, such as the Maya and various Native American tribes, possessed meticulous astronomical knowledge. They built elaborate observatories and monuments aligned with celestial events, reflecting their deep spiritual connection to the cosmos and their reverence for celestial bodies.

These examples merely scratch the surface of the diverse cultural significance of astronomy throughout history. It highlights the universal human curiosity about the stars and the profound impact celestial observations have had on societies across the globe.

In the following chapters of this book, we will continue exploring astronomy's wonders, building upon the rich historical legacy left by these cultures. We will uncover more incredible facts about the cosmos, expanding our understanding of the universe's intricacies and our place within it.

Chapter 2:
Galaxies - Vast Cosmic Islands

Definition and Components of Galaxies

Step into the captivating realm of galaxies, the majestic cosmic constellations scattered throughout the vast expanse of the universe. Within this chapter, we embark on a journey to comprehend the essence of galaxies and immerse ourselves in their enchanting components, unveiling the mysteries and marvels that lie within these celestial wonders.

Galaxies are vast systems of stars, stellar remnants, interstellar gas, dust, and dark matter, all intricately bound together by the force of gravity. These cosmic entities serve as the universe's building blocks, each with its unique structure, properties, and cosmic story.

Within galaxies, stars shine as celestial beacons, illuminating the darkness of space. These stars vary in size, mass, and temperature, forming the mesmerizing mosaic that adorns the galaxies we observe. Galaxies are embellished with a kaleidoscope of solar wonders, from newborn stars nestled within stellar nurseries to ancient giants that have burned bright for eons.

Stellar remnants, such as white dwarfs, neutron stars, and black holes, further enrich the fancy stellar networks of galaxies. These lingering remnants stand as testaments to the celestial bodies that have concluded their life cycles, leaving behind traces of their once-majestic existence. They hold the key to understanding the dramatic processes that occur within galaxies, from stellar birth to explosive deaths.

Interstellar gas and dust pervade galaxies, creating the cosmic nurseries where new stars are born. These molecular clouds and nebulae provide the raw materials for forming stars and planetary systems. They shimmer with vibrant colors and intricate structures, showcasing the dynamic interplay between gravity, radiation, and interstellar matter.

Dark matter, a mysterious substance that eludes direct detection, plays a significant role in the dynamics of galaxies. Its gravitational influence shapes the structure and behavior of galaxies, holding them together and driving their elaborate dance across the cosmic stage. Dark matter remains an question, yet its presence serves as a reminder of the hidden depths that pervade the cosmos.

As we explore the components of galaxies, we embark on a journey through the intricate web of stellar systems that populate the universe. Together, we will uncover the mind-boggling diversity, immense scales, and awe-inspiring beauty that galaxies possess. So, brace yourself as we venture further into the heart of these cosmic islands, unlocking the secrets they hold and expanding our understanding of the cosmos.

Overview of the Milky Way Galaxy and its Significance

Amidst the vast array of galaxies scattered throughout the cosmos, there is one that holds exceptional importance for us: the Milky Way. Within this segment, we embark on a captivating odyssey through our very own galactic abode, revealing its mesmerizing characteristics and delving into its profound relationship with our existence.

The Milky Way is a barred spiral galaxy characterized by its distinct spiral arms that wrap around a central bar-shaped structure. Our Solar System, with its planets, moons, and countless other celestial objects, resides within the expansive boundaries of this remarkable cosmic entity.

Stretching across a staggering diameter of at least 26,800 parsecs (87,400 light-years), the Milky Way boasts a dazzling array of celestial wonders. Its spiral arms house vast collections of stars, interstellar gas, and dust, creating regions of intense stellar formation and cosmic beauty. Stellar clusters, nebulae, and stellar nurseries within the Milky Way serve as celestial cradles for the birth and evolution of new stars.

At the heart of the Milky Way lies a supermassive black hole known as Sagittarius A-star. With a mass equivalent to four million times that of our Sun, this gravitational massive entity influences the surrounding stars, contributing to our astrophysical home's compounded dynamics and structure.

The Milky Way's significance to humanity extends beyond its physical attributes. Throughout history, cultures across the globe have gazed up at the night sky, marveling at the band of diffuse light that spans across the heavens. This ethereal band, known as the Milky Way, has inspired myths, legends, and artistic expressions, connecting humanity to the vastness of the cosmos.

Over the past few centuries, our comprehension of the Milky Way has undergone a remarkable transformation. From initial astronomical observations to the advent of advanced telescopes and space missions, we have progressively revealed the intricate intricacies and profound essence of our universe. The Milky Way serves as a captivating laboratory for investigating the evolution of galaxies, offering valuable insights into the life cycles of stars, the creation of stellar clusters, and the dynamics governing astral structures.

Galactic Halo and Dark Matter

Beyond the visible disk of the Milky Way, an extended region called the galactic halo surrounds the galaxy. The galactic halo is believed to contain significant amounts of dark matter. This invisible substance exerts a gravitational influence on visible matter. Understanding the distribution and properties of dark matter within the Milky Way's halo sheds light on the nature of this elusive cosmic component.

As we traverse the Milky Way's cosmic arms, we deepen our appreciation for the sophisticated patterns of galaxies that decorate the universe. Exploring our cosmic residence sparks curiosity about the countless other galaxies that await our discovery. Together, let us continue our cosmic voyage, venturing into the wonders of the universe and expanding our knowledge of the breathtaking galaxies that fill the cosmic ocean.

Different Types of Galaxies and Their Visual Morphology

Within the expansive interstellar collage of the universe, galaxies come in a mesmerizing array of shapes, sizes, and structures. In this section, we will explore the diverse types of galaxies and delve into their visual morphology, uncovering the remarkable variety among these cosmic entities.

Galaxies are classified into different types based on their visual appearance and characteristics. The three main types of galaxies are elliptical, spiral, and irregular, each offering a unique window into the cosmic wonders of the universe.

Elliptical galaxies are characterized by their rounded or elongated shape, resembling an ellipse. They range in size from small to enormous and contain a diverse population of stars. Elliptical galaxies often lack prominent features such as spiral arms, and their stars tend to follow more random and chaotic orbits within the universe. These galaxies exhibit a remarkable diversity in size, shape, and stellar content, offering a captivating glimpse into the vast cosmic landscape.

On the other hand, Spiral galaxies showcase majestic arms that sweep outward from a central bulge. These arms consist of star clusters, interstellar gas, and dust, creating breathtaking cosmic spirals. Spiral galaxies possess a flattened disk-like structure, with their stars and interstellar matter revolving around a central core. The Milky Way, our own stellar home, is a prime example of a spiral galaxy, as are the stunning Whirlpool Galaxy and the iconic Andromeda Galaxy.

Irregular galaxies defy the conventional patterns observed in elliptical and spiral galaxies. They lack a distinct shape or structure, exhibiting a more chaotic and distinctive appearance. Irregular galaxies can arise from gravitational interactions and collisions between galaxies, resulting in their unique and captivating forms. These galaxies often harbor regions of intense star formation and serve as cosmic laboratories for studying stellar birth and evolution.

Beyond these primary categories, transitional forms and peculiar galaxies defy easy classification, further adding to the richness and complexity of astronomic diversity.

The study of celestial morphology provides us with insights into galaxies' formation, evolution, and dynamics. Astronomers unravel the detailed interplay of gravity, gas dynamics, and stellar processes that shape the cosmos by analyzing their shapes, structures, and composition.

As we explore the vast cosmic ocean, we encounter galaxies of all shapes and sizes, each with its unique story and cosmic legacy. The visual morphology of galaxies beckons us to clarify their mysteries, inspiring awe and wonder at the complex beauty of the universe.

Exhilarating Facts About Galaxies and Their Supermassive Black Holes

As we continue our exploration of galaxies, let us now delve into some stunning facts that deepen our appreciation for these cosmic entities and the extraordinary phenomena they harbor.

Did you know the average galaxy is estimated to contain around 100 million stars? Just imagine the sheer magnitude of stellar systems within a single universe, each twinkling light representing a sun-like entity or a fiery celestial giant. These interstellar realms are veritable galaxies within themselves, hosting stellar communities that dazzle the imagination.

However, the size of galaxies spans an astonishing range. From dwarf galaxies containing fewer than a hundred million stars to the largest known supergiants with a mind-boggling one hundred trillion stars, galaxies demonstrate the immense scale of the cosmos. Each star's unique characteristics and place in the cosmic composition contribute to the vibrant symphony of cosmic life.

A significant portion of a typical galaxy's mass, including our own Milky Way, exists in the form of dark matter. Although invisible to our instruments, this mysterious substance exerts a gravitational force that shapes the dynamics and structure of galaxies. The enigma of dark matter intrigues astronomers and fuels ongoing research into understanding its true nature and influence.

One of the most captivating features at the heart of many galaxies is the presence of supermassive black holes. These gravitational monsters, with masses millions or even billions of times that of our Sun, reside in galactic cores. They draw matter into their gravitational embrace, forming accretion disks that generate intense radiation and jets of high-energy particles. Supermassive black holes play a crucial role in regulating the growth and evolution of galaxies, shaping their structures, and influencing the fate of surrounding stars.

We find the supermassive black hole Sagittarius within our Milky Way galaxy. With a mass four million times greater than our Sun, this cosmic titan captivates the imagination, inspiring a deeper understanding of the intricate dynamics and cosmic forces that mold galaxies.

These mind-blowing facts remind us of the cosmic wonders that galaxies hold. From the sheer scale and diversity of stars within a galaxy to the gravitational enchantment of dark matter and the cosmic influence of supermassive black holes, galaxies continue to captivate and challenge our understanding of the universe.

As we venture further into the realms of galaxies, our journey takes us closer to unveiling the elements that shape the cosmos. Together, let us embrace the wonders of galaxies and expand our knowledge of these magnificent cosmic islands that grace the vast expanse of space.

Chapter 3:
Exploring the Universe - From Dwarfs to Supergiants

Description of the Size Range of Galaxies, From Dwarfs to Supergiants

In our journey through the cosmos, we encounter galaxies of all sizes, spanning a vast range that stretches the limits of our imagination. In this section, we will explore the size spectrum of galaxies, from the dwarfs to the colossal supergiants, revealing the extraordinary diversity within space.

Dwarf galaxies, as their name suggests, are relatively small and compact compared to their larger counterparts. These cosmological entities typically contain fewer than a hundred million stars, making them significantly smaller in scale. However, don't let their size deceive you. Dwarf galaxies can exhibit fascinating features and harbor remarkable phenomena, from intense star formation regions to intriguing interactions with neighboring galaxies.

Moving up the size scale, we encounter galaxies of intermediate size, possessing a moderate number of stars and a more substantial cosmic footprint. These galaxies, often referred to as intermediate-sized or medium-sized galaxies, offer a bridge between the dwarf and supergiant galaxies. Their properties, structure, and evolution display various characteristics that contribute to the complexity of the stellar landscape.

Supergiant galaxies, on the other hand, are the cosmic Goliath of the universe. These colossal entities boast a mind-boggling number of stars, with estimates reaching a staggering one hundred trillion or more. Each supergiant galaxy is a cosmic metropolis, hosting a bustling population of stars, interstellar gas, and celestial formations on an unprecedented scale. The dynamic interactions and gravitational dances within supergiant galaxies shape their systems, driving the evolution and destiny of their stellar residents.

It is within this vast array of galaxies, from dwarfs to supergiants, that we witness the dynamic and ever-evolving nature of the universe. The size of a galaxy not only influences its individual characteristics but also plays a role in its interactions with neighboring galaxies, shaping the cosmic astral composition we observe.

As we navigate the cosmic ocean, from the compact realms of dwarf galaxies to the expansive realms of supergiant galaxies, we gain a deeper appreciation for the breadth and complexity of the astrophysical landscape. Each universe, regardless of its size, holds within it a treasure trove of cosmic wonders and mysteries waiting to be discovered.

Facts About The Immense Number of Stars in Galaxies

Within the vast expanse of galaxies, an astonishing fact awaits the sheer magnitude of stars that populate these cosmic entities. Let us delve into some mind-boggling facts about the immense number of stars within galaxies, shedding light on the extraordinary cosmic tapestry that spans the universe.

When we contemplate the night sky and witness the glittering stars above, it is hard to fathom the true scale of stellar populations within galaxies. On average, a galaxy contains approximately 100 million stars, each a radiant beacon in the cosmic darkness. Just imagine the sheer magnitude of these stellar systems, illuminating the galactic realms with their brilliance.

However, the number of stars within a galaxy can vary significantly. Dwarf galaxies, although smaller in size, still boast a remarkable number of stars, ranging in the millions or tens of millions. The larger and more massive galaxies, such as supergiants, are home to an awe-inspiring one hundred trillion or more stars, painting a cosmic canvas of epic proportions.

Consider the implications of these numbers. The vastness of the universe, with its billions of galaxies, implies an astronomical number of stars. To comprehend the true scope of this, imagine the grains of sand on Earth's beaches, each representing a single star. Now multiply that by billions upon billions, and you might begin to grasp the vastness of the stellar population inhabiting the cosmos.

These staggering numbers highlight the diversity and abundance of stars that make up galaxies. Each star has unique characteristics, from size and temperature to luminosity and lifespan. The interplay between these stars, their gravitational interactions, and their cosmic journeys shape the evolution and dynamics of galaxies themselves.

The realization that galaxies contain such an immense number of stars invites us to ponder our place in the universe. Amidst this vast cosmic sea, our tiny blue planet is a minuscule speck orbiting an ordinary star within the boundaries of a single galaxy.

As we persist in our voyage amid the celestial marvels of the cosmos, let's appreciate the jaw-dropping magnitude and variety of stars nestled in galaxies. Acting as astral lighthouses, they direct our expedition and fuel our fascination with the riddles concealed in the far reaches. In unison, let's be awestruck by the vastness of the star clusters, setting forth on an interstellar journey that broadens our comprehension of the infinite stars lighting up the expansive canvas of the universe.

The Oldest and Most Distant Galaxy Observed, GN-z11.

In the vast cosmic expanse, as we gaze into the depths of space, our eyes are drawn to a captivating celestial entity: GN-z11, the oldest and most distant galaxy ever observed. In this section, we will embark on a journey to uncover the remarkable story of GN-z11, a window into the early epochs of the universe.

GN-z11 is a testament to the incredible power of astronomical observation and technological advancements. Located at an astonishing comoving distance of 32 billion light-years from Earth, GN-z11 allows us to witness the universe as it existed a mere 400 million years after the Big Bang.

The journey to discover GN-z11 was a challenging feat. It required the combined efforts of cutting-edge telescopes and instruments, including the Hubble Space Telescope and the powerful Keck Observatory. By capturing and analyzing the faintest traces of light emitted by this ancient galaxy, astronomers could discover its secrets and gain insights into the early stages of cosmic evolution. It serves as a cosmic time capsule, preserving a snapshot of the universe's infancy and helping us piece together its narrative.

What makes GN-z11 truly exceptional is its age. As we peer into the depths of space and time, we observe this ancient starlit outpost, a relic from the dawn of the universe. Its existence challenges our understanding of cosmic formation and pushes the boundaries of what we thought was possible.

The discovery of GN-z11 is a testament to humanity's insatiable curiosity and our unyielding quest to discover the perplexities of the cosmos. It invites us to contemplate the vastness of space, the passage of time, and the wondrous events that unfold within its embrace.

As we continue explore the marvels of the universe, GN-z11 beckons us to push the boundaries of our knowledge, venture further into the cosmic abyss, and continue seeking answers to the fundamental questions that have captivated humanity for centuries.

Together, let us marvel at the astonishing nature of GN-z11, a distant sentinel from the cosmic past. Its discovery opens new frontiers in our understanding of the universe. It reminds us of the boundless wonders that await our exploration.

Chapter 4:
Dark Matter and Cosmic Mysteries

Overview of the Nature of Dark Matter and its Significance in the Universe

Within the profound reaches of the cosmos, concealed from our sight, dwells a perplexing celestial conundrum: dark matter. In this section, we set forth on an expedition to decipher the secrets surrounding dark matter and examine its pivotal part in the genesis, behaviour, and transformation of galaxies.

Dark matter, as its name suggests, is a mysterious form of matter that does not interact with light or other forms of electromagnetic radiation. It eludes direct detection, rendering it invisible to our instruments and challenging our understanding of the universe. Yet, its gravitational influence permeates the cosmic landscape, shaping the structures and behaviors of galaxies.

The presence of dark matter within galaxies is inferred through its gravitational effects on visible matter. As astronomers study the motions of stars and gas within galaxies, they notice peculiarities that cannot be explained solely by the visible matter we observe. These anomalies suggest the presence of an additional mass in the form of dark matter.

Dark matter is significant in outer space, providing the gravitational glue that holds galaxies together. Its gravitational pull shapes the distribution of matter within galaxies, influencing the rotation curves of stars and determining the structures of spiral arms. Without dark matter, galaxies as we know them would not exist in their current form.

The exact nature of dark matter remains a mystery, leaving scientists with many unanswered questions. Numerous theories have been proposed, suggesting that dark matter may consist of yet-undiscovered subatomic particles that interact only weakly with ordinary matter. Efforts to detect and understand dark matter are ongoing, pushing the boundaries of scientific knowledge and inspiring new avenues of research.

Dark matter's impact extends beyond individual galaxies. It influences the universe's large-scale structure, contributing to the formation of vast cosmic web-like filaments and clusters of galaxies. These cosmic structures, guided by the gravitational embrace of dark matter, shape the distribution of galaxies on the largest scales.

The study of dark matter not only reveal the aspects of astronomic dynamics but also deepens our understanding of the universe's composition. It prompts us to reconsider the nature of matter itself. It challenges our assumptions about the fundamental building blocks of the cosmos.

As we plunge deeper into the celestial void, let's embrace the riddle that is dark matter, permitting its mystery to spark our inquisitiveness and propel our pursuit of understanding. We will investigate dark matter's part in the astral dance of galaxies, deciphering the concealed secrets within the unseen reaches of the cosmos.

Facts About the Mysteries Surrounding Dark Matter

Fact 1: Invisible and Unseen

Dark matter remains invisible and undetectable through conventional means. Despite its substantial presence in the universe, its elusive nature poses a tremendous challenge for scientists seeking direct detection. Its interactions with ordinary matter are fragile, making it difficult to observe and study in detail.

Fact 2: Cosmic Dominance

Dark matter constitutes a staggering 85% of the total matter in the universe, outweighing ordinary matter by a substantial margin. While stars, galaxies, and planets comprise only a small fraction of the cosmic inventory, dark matter holds sway over the cosmic landscape, shaping its structures and dynamics.

Fact 3: Missing Pieces in the Puzzle

The true nature of dark matter remains unknown, representing one of the most compelling puzzles in modern science. Despite decades of research and numerous experiments, scientists have yet to identify the specific particle or particles that comprise dark matter. Its composition and properties remain shrouded in mystery, leaving much to be explored and understood.

Fact 4: Gravitational Influence

Dark matter's gravitational influence extends far beyond its invisible cloak. It plays a crucial role in the formation and dynamics of galaxies, providing the gravitational scaffolding upon which they assemble and evolve. Without dark matter's gravitational pull, galaxies would not have the stability and coherence required to maintain their structures and preserve their stellar populations.

Fact 5: Cosmic Webs and Filaments

The cosmic web, composed of vast filaments and voids that crisscross the universe, owes its existence to dark matter. This multifaceted cosmic tapestry weaves through space, forming a framework for creating galaxy clusters and superclusters. Dark matter's gravitational influence draws ordinary matter along these cosmic highways, sculpting the vast cosmic landscape we observe.

Fact 6: Dark Matter and the Fate of the Universe

Dark matter's significance extends beyond individual galaxies. Its gravitational effects influence the expansion rate of the universe, playing a role in shaping its ultimate destiny. Understanding dark matter is integral to revealing the intricate interplay between dark energy, dark matter, and ordinary matter, collectively determining the fate of the cosmos.

Fact 7: Inspiring New Theories and Experiments

The intricacies surrounding dark matter inspire scientists to develop innovative theories and experiments. From underground detectors to space-based observatories, researchers strive to unlock the secrets of dark matter. The search for dark matter particles continues, pushing the boundaries of our knowledge and opening new avenues for discovery.

These facts highlight the vastness of the cosmic enigma that is dark matter. As we journey through the universe, let us embrace the challenge of unraveling its mysteries, fueled by the curiosity to comprehend the invisible forces that shape the cosmos. Together, we embark on a cosmic quest, pursuing the answers within the shadows of the universe.

Chapter 5:
Astounding Phenomena in Space

As we voyage through the cosmos, we encounter many extraordinary phenomena that captivate our imagination and challenge our understanding of the universe. In this chapter, we will embark on a breathtaking journey to explore some of the most astounding activities that occur within the vastness of space.

Supernova Explosions

Supernovae, the explosive deaths of massive stars, unleash incredible energy and light into the cosmos. These cataclysmic events mark the end of a stellar life cycle and leave behind remnants such as pulsars and neutron stars. The brilliant display of a supernova explosion illuminates the surrounding space, enriching the cosmos with heavy elements like gold, silver, and uranium, scattering them across space, enriching future generations of stars and planetary systems. The components are forged within the heart of the star.

Gamma-Ray Bursts

Gamma-ray bursts are among the most energetic events known in the universe. These intense bursts of gamma-ray radiation result from the collapse of massive stars or the merger of binary systems, releasing enormous amounts of energy across the electromagnetic spectrum. This burst can last from a fraction of a second to several minutes. Gamma-ray bursts offer a glimpse into the extreme physics of the universe and serve as beacons of cosmic evolution. The energy released during a gamma-ray explosion is mind-boggling, often surpassing the total energy output of billions of stars combined.

Quasars

Quasars, short for "quasi-stellar radio sources," are cosmic powerhouses in distant galaxies' hearts. They are fueled by supermassive black holes devouring surrounding matter, emitting intense radiation across various wavelengths. Quasars serve as cosmic lighthouses, illuminating the early universe and providing insights into galactic evolution and the growth of massive black holes. The light we observe from quasars has traveled billions of light-years to reach us, providing a glimpse into the universe's early stages. Some quasars are so far away that their light reveals the universe as it existed when it was only a fraction of its current age.

Blazars

Blazars are a subcategory of quasars characterized by their intense and variable emission of high-energy gamma rays. They are powered by supermassive black holes with powerful jets of particles pointed directly at Earth. Blazars exhibit extreme brightness and serve as natural laboratories for studying the most energetic processes in the universe. These jets, traveling near the speed of light, produce a range of radiation that includes gamma-rays, X-rays, and radio waves. Blazars exhibit extreme brightness and variability, making them fascinating cosmic objects to study.

Pulsars

Pulsars are rapidly rotating neutron stars emitting radiation beams that sweep across space like cosmic lighthouses, the remnants of massive stars that have undergone supernova explosions. They emit beams of radiation that sweep across space, resulting in regular pulses of emission observed from Earth. Pulsars can rotate hundreds of times per second, and their intense magnetic fields focus and accelerate particles. Their highly regular pulses are scanned across the electromagnetic spectrum, ranging from radio waves to X-rays and gamma rays. Pulsars offer insights into the nature of matter under extreme conditions and provide a unique window into the remnants of stellar explosions. Some pulsars even serve as highly accurate cosmic clocks, rivaling Earth's most precise atomic clocks.

Cosmic Microwave Background Radiation

The cosmic microwave background (CMB) radiation is the faint echo of the universe's early moments, dating back to about 380,000 years, known as the "afterglow" of the Big Bang. It consists of faint microwave radiation that permeates the entire universe. The CMB is often described as the universe's first light, representing the moment photons could finally travel freely through space. It pervades the whole universe, serving as a relic from the hot, dense phase of cosmic infancy. The CMB provides valuable clues about the composition, age, and expansion of the universe, offering a glimpse into our cosmic origins.

These truths augment our admiration for the extraordinary celestial events transpiring within the universe. They serve as reminders of the immense energy, extensive spans, and profound cosmic mechanisms.

Chapter 6:
Unveiling the Local Group: A Galactic Ensemble of Marvels and Interactions

This section will broaden our cosmic perspective beyond the Milky Way and explore the Local Group. This small interstellar neighborhood encompasses our own galaxy and the neighboring Andromeda Galaxy. Join us as we uncover this cosmological ensemble, fascinating details and remarkable interactions.

The Local Group

The Local Group is a small cluster of galaxies that includes approximately 54 galaxies, gravitationally bound together in a cosmic dance. At the heart of this group lies our very own Milky Way, along with our closest celestial neighbor, the Andromeda Galaxy (M31), and a collection of smaller dwarf galaxies. The Local Group is a testament to the thorough interplay of gravitational forces and cosmic dynamics on a small scale.

The Andromeda Galaxy (M31)

The Andromeda Galaxy, also known as M31, is prominent in the Local Group. Located about 2.537 million light-years from the Milky Way, it is the nearest spiral galaxy to us. With a diameter of approximately 220,000 light-years, the Andromeda Galaxy is slightly larger than the Milky Way and contains a staggering trillion stars. Its majestic spiral arms, dust lanes, and stellar clusters make it a captivating sight in the night sky.

Galactic Collisions

One of the most fascinating aspects of the Local Group is the future cosmic collision between the Milky Way and the Andromeda Galaxy. Due to their gravitational attraction, the two galaxies are on a collision course, culminating in a astronomic merger billions of years from now. This collision will reshape the structure of both galaxies, triggering star formation and leaving a new, combined starlit entity in its wake.

Satellite Galaxies

The Local Group is not solely composed of the Milky Way and the Andromeda Galaxy. It also hosts numerous satellite galaxies, smaller dwarf galaxies that orbit around the more massive galaxies. Examples include the Large and Small Magellanic Clouds, satellite galaxies of the Milky Way, and the Triangulum Galaxy (M33), a companion of the Andromeda Galaxy. These satellite galaxies contribute to the perplexing dynamics and gravitational interactions within the Local Group.

Gravitational Dance

Within the Local Group, the gravitational interactions between galaxies shape their orbits and influence their evolution. Galaxies are not static entities but are constantly in motion and interaction. The delicate balance of gravitational forces determines galaxies' paths, leading to detailed patterns of orbits and alignments within the Local Group.

Exploring the Local Group and the Andromeda Galaxy allows us to broaden our perspective and appreciate the interconnectedness of the cosmic landscape. As we witness the cosmic interactions and foresee the future merger between the Milky Way and Andromeda, we gain insights into our galactic neighborhood's cosmic evolution and dynamic nature.

Together, let us continue our cosmic exploration, embracing the Local Group's wonders and the Andromeda Galaxy's mesmerizing beauty. As we unveil the secrets of this galactic ensemble, we deepen our understanding of the cosmic symphony that unfolds in the vastness of the universe.

Facts About the Vast Cosmic Structures: Sheets, Filaments, and Voids

This section delves into the astral patterns extending far beyond individual galaxies—sheets, filaments, and voids. Brace yourself for mind-blowing facts that shed light on these cosmic landscapes' immense scales and breathtaking beauty.

Cosmic Web

The universe is not a randomly scattered collection of galaxies. Instead, it forms a delicate cosmic web—a vast network of interconnected structures. This web comprises cosmic sheets, filaments, and voids, weaving a mesmerizing tapestry across the cosmos. The cosmic web represents the large-scale distribution of matter, shaped by the gravitational pull of dark matter.

Sheets

Cosmic sheets are immense, flat regions of the universe, stretching across millions of light-years. These expansive sheets are composed of clusters, superclusters, and groups of galaxies meticulously arranged in a filamentary structure. Sheets are like cosmic walls, forming boundaries that define the vast voids between them.

Filaments

Filaments are elongated threads of matter that connect cosmic sheets. They act as bridges, linking galaxies and galaxy clusters. Filaments span hundreds of millions of light-years, elaborately interweaving through the cosmic web. These filamentary structures are rich in matter, forming the backbone of the cosmic web's refined architecture.

Voids

Voids are vast, seemingly empty regions within the cosmic web. These expansive regions stretch between filaments and sheets, creating cosmic voids of unimaginable proportions. Voids can be hundreds of millions of light-years wide, appearing as vast cosmic bubbles. While voids appear empty, they contain traces of matter, including dwarf galaxies, cosmic dust, and sparse gas.

Great Attractor

Within the cosmic web, a gravitational anomaly, the Great Attractor, exerts its pull. The Great Attractor is a region of space that influences the motion of galaxies, causing them to move towards it. This mysterious cosmic structure, hidden behind the Milky Way's center, remains partially obscured. Demystifying the nature and origin of the Great Attractor is a challenge that intrigues scientists.

Revealing the Universe's Grand Design

The study of the cosmic web offers a glimpse into the underlying structure of the universe, revealing the interplay between dark matter, ordinary matter, and cosmic evolution. Through advanced observational techniques and computer simulations, scientists have mapped the detailed patterns of the cosmic web, unmasking the secrets of the universe's large-scale architecture.

Laniakea Supercluster

The Local Group, including the Milky Way and the Andromeda Galaxy, is part of a larger cosmic structure known as the Laniakea Supercluster. Laniakea spans over 500 million light-years and encompasses thousands of galaxies. Within Laniakea, the gravitational interactions of galaxies shape the cosmic flows and guide the motion of astral ensembles.

These facts highlight the awe-inspiring cosmic structures that extend across vast distances. The sheets, filaments, and voids of the cosmic web shape the grand cosmic symphony, influencing galaxies' formation and driving the universe's evolution.

As we contemplate the vastness of the cosmic web, let us be humbled by the puzzling masterpiece that connects galaxies across billions of light-years. The cosmic structures that span the universe remind us of the extraordinary scales at play and inspire us to continue exploring the wonders within the boundless depths of the cosmos.

Chapter 7:
Cosmology - The Study of the Universe

In this chapter, we embark on a profound exploration of cosmology, the scientific discipline dedicated to understanding the universe. Then, we will delve into the fundamental concepts and captivating discoveries that define cosmology as a branch of astronomy.

Definition of Cosmology

Cosmology is the scientific study of the universe's origin, evolution, structure, and ultimate fate. It seeks to explain the oddities of the cosmos by investigating the fundamental questions about the nature of space, time, matter, and energy on the largest scales. Cosmology combines observations, theoretical models, and mathematical frameworks to deepen our understanding of the universe's grand design.

The Big Bang Theory

At the heart of cosmology lies the Big Bang theory—a widely accepted model that describes the universe's birth. According to this theory, the universe originated from a sweltering and dense state approximately 13.8 billion years ago. The subsequent expansion of space-time gave rise to the universe as we know it today. The Big Bang theory provides a framework for understanding the origin and early evolution of the cosmos.

Cosmic Microwave Background

As mentioned previously, the cosmic microwave background (CMB) radiation plays a pivotal role in cosmology. It is the faint radiation left over from the hot, dense phase of the early universe when it transitioned from a plasma to a transparent state. The study of the CMB allows cosmologists to probe the universe's early conditions, test the predictions of the Big Bang theory, and gain insights into the composition and evolution of the universe.

Large-Scale Structure

Cosmologists investigate the universe's large-scale structure, mapping the distribution of galaxies, galaxy clusters, and cosmic voids. By studying these structures, cosmologists gain valuable information about the underlying cosmic web and the processes that shaped the universe over billions of years. Studying large-scale networks provides insights into cosmic expansion, the influence of dark matter and energy, and the formation of cosmic filaments and superclusters.

Dark Matter and Dark Energy

Cosmology also grapples with the mysteries of dark matter and dark energy. Dark matter, which outweighs visible matter in the universe, plays a crucial role in the formation of structures and the gravitational dynamics of galaxies. On the other hand, dark energy is a mysterious force driving the universe's accelerated expansion. Understanding the nature and properties of dark matter and energy are among cosmological research's central pursuits.

Multiverse and Beyond

Cosmology also explores the possibility of a multiverse—an ensemble of multiple universes beyond our observable realm. The concept of a multiverse arises from theoretical models, such as inflationary cosmology, and the search for a deeper understanding of the universe's fundamental laws and parameters. Investigating the concept of a multiverse challenges our notions of reality and pushes the boundaries of cosmological exploration.

Cosmology stands at the forefront of scientific inquiry, striving to uncover the fundamental truths and mysteries that shape our cosmic existence. By peering into the vast depths of space and time, cosmologists provide us with insights into the universe, its origins, its evolution, and its destiny.

Facts about the Big Bang and Cosmic Background Radiation

This section reveals facts illuminating the nature of the Big Bang and the cosmic background radiation. Prepare to be amazed as we explore the remarkable aspects of these fundamental cosmic observations.

The Primordial Fireball

The Big Bang was not an explosion in space but rather the sudden emergence of space and time. In the tiniest fraction of a second, the universe expanded from an incredibly hot and dense singularity, where all matter and energy were compressed into an infinitesimally small point. This primordial fireball marked the beginning of our cosmic journey.

Rapid Expansion

The expansion of the universe during the Big Bang was extraordinarily rapid. In a mere fraction of a second, space expanded exponentially, stretching the fabric of the universe to unimaginable scales. This rapid expansion, known as cosmic inflation, smoothed out irregularities and set the stage for forming galaxies and cosmic structures.

The First Light

As mentioned in the previous chapter, approximately 380,000 years after the Big Bang, the universe had cooled enough for protons and electrons to combine and form neutral atoms. This influential moment, recombination, allowed photons to travel freely through space. These photons, which comprise the cosmic microwave background radiation, carry information about the universe's early conditions and give us a glimpse into its infancy.

The Oldest Light

The cosmic microwave background radiation is often described as the oldest light in the universe. It has traveled through space for over 13 billion years, echoing the universe's early moments. By studying the patterns and fluctuations in the CMB, cosmologists can glean insights into the composition, geometry, and evolution of the universe.

Striking Uniformity

One of the most astonishing facts about cosmic background radiation is its remarkable uniformity. The temperature of the CMB is almost identical in all directions of the sky, with only slight temperature variations of a few parts in a million. This astonishing uniformity challenges our understanding of how such uniformity emerged from the fiery chaos of the early universe.

Nobel Prize Discovery

The discovery of cosmic microwave background radiation is considered one of the greatest achievements in cosmology. In 1964, Arno Penzias and Robert Wilson accidentally stumbled upon the CMB while investigating radio waves. Their groundbreaking discovery earned them the Nobel Prize in Physics in 1978, solidifying the Big Bang theory and paving the way for our understanding of the universe's origins.

The Big Bang and the cosmic background radiation captivate us with their intriguing aspects and offer a glimpse into the earliest moments of cosmic history. They remind us of the incredible journey the universe has taken from a primordial singularity to the vast cosmos we behold today.

As we ponder the significance of the Big Bang and the cosmic background radiation, we are humbled by the intricate cosmic dance that has shaped our existence. Let us continue to unravel the anomalies of the universe, exploring the depths of cosmic time and space and marveling at the astonishing wonders that await our discovery.

Chapter 8:
Astronomy vs. Astrophysics

Differentiating Between Astronomy and Astrophysics

In this chapter, we dive into the fascinating realms of astronomy and astrophysics, exploring these two disciplines' similarities, distinctions, and interconnected nature. Then, we will navigate the celestial landscapes and uncover the nuances that set astronomy and astrophysics apart.

Astronomy

Astronomy is an ancient science that focuses on the observation and study of celestial objects. It encompasses the exploration of planets, moons, stars, galaxies, nebulae, and other cosmic entities. Astronomers use telescopes and other observational tools to collect data about celestial objects' properties, behavior, and interactions. Astronomy is often described as the qualitative study of the universe, seeking to understand its vastness and beauty.

Astrophysics

Astrophysics, on the other hand, is a branch of astronomy that delves into the underlying physics and mathematical principles governing celestial objects' behavior, composition, and evolution. It applies the laws of physics, such as gravity, electromagnetism, and thermodynamics, to study the physical properties and processes occurring in the cosmos. Astrophysicists employ mathematical models, simulations, and advanced instruments to dissect the intricate mechanisms in stars, galaxies, black holes, and other cosmic spectacles. Astrophysics is often considered the quantitative and theoretical counterpart to astronomy.

Interconnectedness

While astronomy and astrophysics are distinct fields, they are intimately connected, with overlapping areas of study. Astronomers and astrophysicists often collaborate and share insights as their research interests and objectives align. Astronomy provides the observational foundation upon which astrophysics builds theoretical models and explanations. Astrophysics, in turn, deepens our understanding of the physical processes underlying the observations made by astronomers. Together, these disciplines contribute to our comprehensive knowledge of the universe.

Historical Perspective

Historically, the boundary between astronomy and astrophysics was less defined. The term "astrophysics" emerged in the late 19th century to denote the physics-based approach to studying celestial phenomena. Before that, "astronomy" and "astrophysics" were often used interchangeably. Over time, as our knowledge of the universe expanded and scientific techniques advanced, the distinction between astronomy and astrophysics became more prominent.

Evolving Fields

Both astronomy and astrophysics are dynamic fields that continuously evolve with technological advancements and scientific understanding. New observational instruments, space missions, and computational tools allow astronomers and astrophysicists to probe deeper into the cosmos, uncovering new developments and addressing previously unanswerable questions. The boundaries between the two disciplines continue to blur as they converge in their pursuit of unraveling the paradoxes of the universe.

Career Paths

Pursuing a career in astronomy or astrophysics can lead to various professional paths. Astronomers often work in observatories, planetariums, or universities, engaging in observational research, public outreach, and teaching. On the other hand, astrophysicists may be found in academic institutions, research facilities, or even interdisciplinary environments, collaborating with physicists, mathematicians, and computer scientists. Both disciplines offer exciting opportunities for those passionate about exploring the cosmos.

Explanation of the Complementary Roles of Observational and Theoretical Astronomy

In this section, we discuss the complementary roles of observational and theoretical astronomy, illuminating how these two approaches work hand in hand to deepen our understanding of the universe.

Observational Astronomy

Observational astronomy focuses on gathering and analyzing data from observations of celestial objects and elements. Astronomers utilize a variety of instruments, such as telescopes, spectrographs, and detectors, to capture light, radio waves, or other forms of electromagnetic radiation emitted by celestial sources. Observational astronomers meticulously analyze these data to uncover patterns, measure properties, and study the behavior of objects across the universe.

Advancements in Observation

Technological advancements have revolutionized observational astronomy, allowing us to probe deeper into the cosmos with increasing precision and sensitivity. Our ability to capture detailed observations has expanded dramatically from ground-based telescopes equipped with adaptive optics to space-based observatories like the Hubble Space Telescope. Observational astronomers study diverse objects, including planets, stars, galaxies, and cosmic anomalies, enabling us to explore the universe across various scales.

Theoretical Astronomy

Theoretical astronomy complements observational astronomy by utilizing mathematical models, simulations, and theoretical frameworks to understand the physical processes occurring in the universe. Theoretical astronomers develop mathematical equations and computational models that describe celestial objects' behavior, formation, and evolution. They use these models to make predictions, test hypotheses, and explain the observational data obtained by astronomers.

Unveiling the Mysteries

Observational astronomy provides empirical evidence and observational constraints that guide the development of theoretical models. Theoretical astronomers, in turn, refine these models to explain the observed events and predict new circumstances that may be discovered through observations. The iterative process between observation and theory drives our understanding of the universe, allowing us to solving its mysteries and push the boundaries of knowledge.

Feedback Loop

The interplay between observation and theory forms a feedback loop that drives scientific progress in astronomy. Observations inspire new theoretical ideas and models, while theories guide the design of observational experiments and the interpretation of data. This feedback loop fosters a deeper understanding of the cosmos, refining our knowledge of celestial objects, their interactions, and the physical laws governing their behavior.

Interdisciplinary Collaborations

The complementary roles of observational and theoretical astronomy often lead to interdisciplinary collaborations. Observational astronomers work closely with theoretical astronomers to refine models, interpret observations, and explore new avenues of research.

Additionally, partnerships with physicists, mathematicians, and computer scientists further enhance our understanding of the universe, combining expertise from different disciplines to tackle complex astrophysical problems.

By adopting both observational and theoretical methodologies, astronomers stretch the limits of our understanding, revealing the secrets of the cosmos. The fusion of tangible data and theoretical structures allows us to gain a holistic comprehension of the universe, tracing its journey from inception to its current condition.

Chapter 9:
Amateur Astronomy - Unveiling the Transient

The Power of Passion

Amateur astronomers are driven by a deep passion for the night sky and a curiosity about the cosmos. With their enthusiasm and dedication, they actively observe and study celestial objects, often pursuing their hobby with advanced equipment and a commitment to expanding their knowledge. This passion fuels their contributions to astronomy and opens new avenues of discovery.

Discovering Transient Events

Transient events are short-lived astronomical events that occur unexpectedly, such as supernovae, comets, meteor showers, and even the occasional outburst of a variable star. Amateur astronomers play a crucial role in discovering and monitoring these transient events. Their vigilance, meticulous observations, and rapid reporting provide valuable data that allows professional astronomers to study these events in real time and gain deeper insights into their nature and characteristics.

Supernova Discoveries

Amateur astronomers have made significant contributions to supernova discoveries over the years. Their keen eye for detail and commitment to regular sky surveys have led to the detection numerous supernovae in galaxies near and far. By identifying these stellar explosions, amateurs contribute to our understanding of stellar evolution, galaxies' dynamics, and the universe's distribution of matter.

Comet Hunting

Amateurs have a long and storied history of discovering comets. With their patient sky scans and astute observations, they have been instrumental in spotting comets as they journey through the solar system. These amateur comet hunters contribute to our knowledge of the composition and behavior of these icy visitors, offering valuable insights into the formation and evolution of our planetary system.

Variable Star Observations

Variable stars undergo periodic changes in brightness due to intrinsic properties or external factors. Amateur astronomers play a vital role in monitoring and studying these variable stars. By carefully observing and recording changes in brightness over time, amateurs contribute to understanding stellar pulsations and binary star systems that shape the life cycles of stars.

Citizen Science and Collaborations

The active involvement of amateur astronomers extends beyond individual observations. Many amateurs participate in citizen science projects and collaborate with professional astronomers. These collaborations enable them to contribute to large-scale research initiatives, such as exoplanet discoveries, light curve analysis, and tracking asteroid orbits. The collective efforts of amateurs and professionals enrich our understanding of the universe and foster a sense of community within the astronomical community.

Outreach and Education

Amateur astronomers also play a crucial role in public outreach and education. Their passion and knowledge inspire others to look at the night sky and develop an interest in astronomy. They share their expertise through public star parties, workshops, and online communities, guiding newcomers and kindling a sense of wonder about the universe.

The active role of amateurs in astronomy is a testament to the inclusive and collaborative nature of the field. Their passion, dedication, and contributions make them indispensable partners in pursuing astronomical knowledge. Together, amateurs and professionals push the boundaries of exploration and inspire future generations to gaze at the stars with curiosity and awe.

Next, celebrate the invaluable contributions of amateur astronomers and explore the transient events that shape our understanding of the ever-changing cosmos. Let us revel in their passion, dedication, and unwavering commitment to unraveling the wonders that unfold in the vast expanse of the universe.

Facts About Significant Discoveries Made by Amateur Astronomers

Comet Hale-Bopp

One of the most famous comet discoveries by amateur astronomers is Comet Hale-Bopp. In 1995, Alan Hale and Thomas Bopp independently spotted the comet, marking one of the most extraordinary comets of the 20th century. Their discovery captured the world's attention. Hale-Bopp went on to provide a spectacular show in the night sky for over a year.

Exoplanets

Amateur astronomers have played a pivotal role in discovering exoplanets, planets orbiting stars beyond our solar system. In 1999, amateur astronomer Bill Doyle detected a transit of the exoplanet HD 209458b, becoming the first amateur to observe a planet crossing in front of its parent star. This groundbreaking observation demonstrated the potential of amateurs in contributing to the growing field of exoplanet research.

Supernova SN 2011fe

In 2011, amateur astronomers Pugh and Bode discovered supernova SN 2011fe in the nearby galaxy M101. Their vigilance and sharp-eyed observations enabled scientists to study this particular supernova, contributing to our understanding of stellar explosions and the formation of heavy elements in the universe.

Variable Stars

Amateurs have made numerous discoveries and observations of variable stars, which undergo periodic changes in brightness. One notable example is the amateur astronomer Clyde Tombaugh, who discovered Pluto in 1930 while searching for a hypothetical ninth planet. Tombaugh's discovery forever changed our understanding of the solar system and opened up new avenues of exploration.

Asteroids and Near-Earth Objects

Amateur astronomers have made substantial contributions to discovering and tracking asteroids and near-Earth objects (NEOs). Their vigilant surveys and systematic observations have resulted in the detection of numerous asteroids, some of which have potentially hazardous orbits. These amateur discoveries contribute to our knowledge of the asteroid population and help safeguard our planet.

Lunar Impact Monitoring

Amateurs have actively monitored the Moon for impact events caused by meteoroids striking its surface. Their observations and recordings of impact flashes have contributed valuable data to lunar science, providing insights into the frequency and dynamics of lunar impacts and the potential risks they pose to future lunar missions.

Conclusion

As we come to the end of our journey through The Abyss Above: Mind-Blowing Facts About Astronomy, the Cosmos, and Outer Space, let us take a moment to reflect on the fascinating facts and topics we have explored:

We started our exploration by diving into the fundamentals of astronomy, understanding celestial objects, their properties, and the methods used to study them.

We marveled at the immensity of galaxies, learning about their vast numbers, sizes, and the unknowns they hold within.

The enigmatic nature of dark matter captivated our attention as we explored its significance in galaxies and the cosmos.

We delved into the enchantments of the oldest and most distant galaxy observed, GN-z11, and pondered the cosmic origins it represents.

The wonders of space phenomena, from supernovae explosions to cosmic background radiation, left us in awe of the universe's dynamic nature.

Our exploration of the Milky Way and beyond revealed the intricacies of our home galaxy and its place within the larger cosmic landscape.

We marveled at the vast cosmic structures, such as sheets, filaments, and voids, that shape the cosmic web and our understanding of the universe's large-scale structure.

The branch of cosmology offered insights into the origin and evolution of the universe, taking us on a journey from the Big Bang to cosmic background radiation.

We distinguished between astronomy and astrophysics, appreciating their complementary roles in interpreting the cosmos.

The active role of amateur astronomers in unveiling transient events and contributing to discoveries and observations fascinated us, highlighting their passion and dedication.

Final thoughts on the Nature of Astronomy and the Cosmos

Astronomy, the study of the universe, continues to astound us with its profoundness, captivating beauty, and limitless potential for discovery. The cosmos, with its billions of galaxies, trillions of stars, and countless celestial experiences, beckons us to explore its depths and expand our understanding of the vast expanse we reside.

Throughout this book, we have been exposed to the wonders of astronomy, from the grandeur of galaxies to the intricacies of subatomic particles. We have witnessed the power of observation and the role of theoretical frameworks in uncovering the secrets of the universe. We have celebrated the contributions of novice astrologists who, with their passion and dedication, have made significant discoveries and expanded our knowledge of the cosmos.

The inspiring nature of astronomy reminds us of our place in the universe. It ignites a sense of wonder within us. It fuels our curiosity, drives scientific progress, and inspires us to reach for the stars in pursuit of knowledge.

As we conclude this book, let us carry the remarkable facts, insights, and stories we have encountered. Let us gaze up at the night sky with newfound appreciation, knowing that each twinkling star holds a story waiting to be told. And let us continue our exploration of the cosmos, for the universe beckons us to unravel its secrets, one astro nugget at a time.

Epilogue

Encouragement for Further Exploration and Learning in Astronomy

Congratulations! You have completed your journey through The Abyss Above: Mind-Blowing Facts About Astronomy, the Cosmos, and Outer Space. But let this not be the end of your astronomical adventure. Instead, let it be the beginning of a lifelong exploration and a catalyst for further learning and discovery.

The universe is an ever-expanding spectrum of knowledge waiting to be discovered. Countless celestial wonders are yet to be explored, ambiguities waiting to be solved, and new discoveries waiting to be made. So, I encourage you to continue your journey of exploration in astronomy, for it is a field that offers endless opportunities to satisfy your curiosity and expand your horizons.

Take the time to gaze at the night sky and observe the beauty of the stars, planets, and galaxies. Seek opportunities to attend stargazing events, join local astronomy clubs, or participate in citizen science projects. Engage in discussions with fellow enthusiasts, share your observations and insights, and be open to learning from others. Remember, astronomy is a field that thrives on collaboration and knowledge sharing.

Most importantly, never stop asking questions. Let your curiosity guide you on a quest for deeper understanding. Explore the latest research, delve into the scientific literature, and stay informed about new discoveries and advancements. The universe is dynamic, and our comprehension of it continues to evolve. You can stay at the forefront of astronomical knowledge by keeping up with the latest developments.

Additional resources for those interested in delving deeper into the subject

Suppose you're eager to delve deeper into the captivating world of astronomy. In that case, numerous resources are available to aid you on your journey. Here are some additional resources that can help expand your knowledge and satisfy your astronomical curiosity:

1. Books:

- "Cosmos" by Carl Sagan
- "A Brief History of Time" by Stephen Hawking
- "The Universe in a Nutshell" by Stephen Hawking
- "The Elegant Universe" by Brian Greene
- "Astrophysics for People in a Hurry" by Neil deGrasse Tyson

2. Websites and Online Resources:

- NASA's website (www.nasa.gov): Explore the latest news, images, and discoveries from space missions and research projects.

- The Hubble Space Telescope website (<u>hubblesite.org</u>): Discover breathtaking images and learn about the latest findings from one of the most iconic telescopes in history.
- The European Space Agency's website (<u>www.esa.int</u>): Access information about European space missions, research, and discoveries.

3. Observatories and Planetariums:

Visit local observatories, planetariums, or science centers in your area. Many of them offer public programs, stargazing events, and educational exhibits.

4. Online Astronomy Courses and MOOCs:

Platforms such as Coursera (<u>www.coursera.org</u>) and edX (<u>www.edx.org</u>) offer online courses in astronomy and astrophysics taught by renowned professors from top universities.

Resources

American Astronomical Society. (n.d.). https://aas.org/

Astronomy. (n.d.). Caltech Astro Outreach. http://www.astro.caltech.edu/

Astronomy magazine. (2023, July 13). Astronomy Magazine - Interactive Star Charts, planets, meteors, comets, telescopes. Astronomy Magazine. https://www.astronomy.com/

European Space Agency. (n.d.). https://www.esa.int/

Florin Simion Yonescat. (n.d.). Royal Astronomical Society. The Royal Astronomical Society. https://ras.ac.uk/

Hubble Home. (n.d.). Hubble. https://hubblesite.org/

Jenner, N. (2023). 20 amazing facts about space and astronomy. www.skyatnightmagazine.com. https://www.skyatnightmagazine.com/space-science/facts-about-astronomy-space/

National Aeronautics and Space Administration. (n.d.). NASA. https://www.nasa.gov/

National Air and Space Museum. https://airandspace.si.edu/

SDSS. (2023, June 7). Sloan Digital Sky Survey-V: Pioneering Panoptic Spectroscopy - SDSS-V. SDSS - Mapping the Universe. https://www.sdss.org/

Space.com: NASA, space exploration and astronomy news. (2023, July 16). Space.com. https://www.space.com/

Space News. (n.d.). National Geographic. https://www.nationalgeographic.com/science/space/

Wikipedia contributors. (2023a). Astronomy. Wikipedia.
https://en.wikipedia.org/wiki/Astronomy

Wikipedia contributors. (2023). Galaxy. Wikipedia.
https://en.wikipedia.org/wiki/Galaxy

The Abyss Below:

Astonishing Ocean Facts and Legends. An In-depth Exploration of Our Planet's Watery Realms

Introduction

Step into the world of wonder and mystery that is our planet's greatest marvel: the ocean. "Into the Abyss: Astonishing Ocean Facts & Legends - An In-depth Exploration of Our Planet's Watery Realms" is more than an ordinary book. It is a journey into the heart of the deep blue, a voyage that promises to captivate, educate, and inspire. It calls to uncover the untamed and unexplored, learn, and dream.

This book unravels the complex tapestry of scientific knowledge, ancient history, and timeless folklore surrounding the world's oceans. It's a deep dive into the grandeur of the oceanic world, from its surface shimmering under the sun's golden touch to its dark abysses, where light fears to tread.

A truly comprehensive resource, this book touches upon the formation of the oceans, their multifaceted roles and functions in our planet's survival, and the dance of life beneath the waves. It is not just a scientific exploration; it is also an adventure into the world of myths and legends, those beguiling tales spun around the ocean's enigmatic depths, capturing the imagination of generations and inspiring countless stories of sea monsters, lost cities, and fabled treasures.

Importance of Oceans in Our World

Imagine a world without oceans. It's hard. That's because oceans are the lifeblood of our planet. The vast blue seas, covering over 70% of

Earth's surface, are the heartbeats of life as we know it. They serve as the largest habitat in the world. In this home, diverse species of all sizes and shapes reside, from the microscopic plankton to the majestic blue whale. Many of these species remain unknown, their secrets safely locked away in unexplored depths.

Oceans are a global thermostat, governing the Earth's climate, influencing weather patterns, and nurturing life on a grand scale. They absorb carbon dioxide, producing oxygen, making every other breath we take a gift from the ocean. They are the veins of our planet, circulating currents of water that transport heat, determining the climate of the world's continents.

They are the arteries of global trade and commerce, pathways for ships carrying goods across nations. Yet, despite their undeniable importance, our understanding of these grand bodies of water is just a drop in the ocean. Our knowledge has yet to plunge into the depths of the seas to uncover the many mysteries within their depths.

This book is a testament to the majesty of our oceans, the very cradle of life. It is a call to every reader to appreciate, learn, and protect our oceans, our planet's most priceless resource, a world within our world that remains largely unexplored and misunderstood. Through its pages, we invite you to embark on this voyage of discovery, unlock the secrets of the ocean, and in the process, discover more about our extraordinary planet and ourselves.

Chapter 1:
An Oceanic Overview

Explanation of Terms: The Ocean and The Sea

As we embark on this oceanic journey, it's vital to grasp the words that serve as our navigational aids. When we speak of 'the ocean' or 'the sea,' we often refer to the vast, interconnected expanse of saltwater that swathes over two-thirds of our planet's surface. This is not a singular, solitary entity but a complex system comprising the Atlantic, Pacific, Indian, Southern, and Arctic Oceans.

These terms, though frequently used interchangeably, do carry nuanced differences. A 'sea' is typically a smaller body of water, partially or entirely encompassed by land. Names like the North, Red, or Mediterranean Sea immediately come to mind. However, the distinction between 'seas' and 'oceans' is not razor-sharp. It's a gradient rather than a strict partition, a spectrum of marine environments defined by their sizes, geographic features, and ecological attributes.

One general rule of thumb is that seas are smaller than oceans and are often partially (as marginal seas) or entirely (as inland seas) bordered by land. So, while every sea is part of the ocean, not every aspect of the ocean is a sea. This distinction may seem subtle, but it's a fundamental part of understanding the complexity and variety within the world's water bodies. As we journey across the world's oceans, this understanding will help us appreciate the diverse landscapes and ecosystems that vary from the open ocean's vast expanses to the sheltered coastlines of our world's seas.

The ocean and sea play more than just a physical role; they also have a cultural one. In British English, for instance, 'the sea' is often the term of choice, irrespective of whether the reference is to a specific ocean. These linguistic peculiarities reflect our human perceptions and interpretations of the natural world and add another layer of richness to our exploration of the watery kingdoms.

Exploration of the World Ocean Concept

Once we've grappled with the seas and oceans' distinctions, we'll explore a concept that reframes our understanding of these bodies of water - the 'World Ocean.' This term, coined in the early 20th century by the visionary Russian oceanographer Yuly Shokalsky, encapsulates the idea of a singular, interconnected body of salt water that covers and embraces most of Earth. The 'World Ocean' is not simply a body of water; it is a living, breathing entity that influences everything from the climate to the air we breathe.

This concept underscores the oceans' vastness and connectivity, emphasizing the seamless interchange among its constituent parts. Understanding our global ocean as an interconnected entity rather than disjointed water sections is fundamental to grasping the dynamics of oceanographic processes. It helps us visualize how currents circulate nutrients, how changes in one area can affect remote parts of the world, and how the oceans collectively moderate global climate patterns.

As we plunge into the ocean's mysteries, we must hold onto this perspective of unity and connectivity. Each wave that crashes onto the shore, every creature we will encounter in the depths, is part of this greater whole of the World Ocean. Recognizing this is significant for scientific understanding; it's a poignant reminder of our shared responsibility. The health of any single part of the ocean impacts the entire body's well-being. Therefore, we must see every action taken toward ocean conservation as a step in safeguarding this invaluable, interconnected ecosystem. As we voyage further 'Into the Abyss,' we carry this collective consciousness, realizing our intrinsic link to the ocean's fate and our vital role in its preservation.

Chapter 2:
The Etymology of 'Ocean'

Historical Origin of the Term 'Ocean'

In Chapter 2, we embark on a linguistic journey, tracing the roots of the term 'ocean' back to ancient times. This journey is not merely an etymological exploration; it also unveils how our ancestors viewed the vast water surrounding them and influenced their civilizations.

The word 'ocean' draws its origin from ancient Greek mythology and language, specifically from the figure in classical antiquity, Oceanus. Recognized as the elder of the Titans in classical Greek mythology, Oceanus personified the enormous river believed to encircle the world. This belief not only influenced the term we use today but also reveals how the ancient Greeks and Romans perceived their world - as an island surrounded by an enormous, all-encompassing body of water.

Moreover, Oceanus was often depicted with a dragon tail, a significant mythological motif in its own right. Dragons, as mythical creatures, are usually associated with the primal forces of nature, embodying power, mystery, and often a certain degree of danger or chaos. By associating the ocean with such a powerful symbol, the ancient Greeks captured its dual nature—its capacity to nurture life and its potential for destruction.

But our exploration continues beyond there. The etymology of the word 'ocean' reveals deeper connections across cultures. The concept of Oceanus has been linked to the Indo-European language family, one of the most widely spoken linguistic groups globally. The Greek' Ōkeanós' has been compared to the Vedic epithet ā-śáyāna-, which was attributed to the dragon Vṛtra-, who captured the cows/rivers in ancient Indian mythology.

This interconnectedness of linguistic and mythological references paints a vivid picture of how the concept of the ocean has transcended cultural boundaries and stood the test of time. This common thread may suggest a shared linguistic ancestry and a universal recognition of the ocean's significance and omnipresence in our lives.

The origin of the word 'ocean' takes us back in time, making us realize that our fascination and reverence for the vast waters are as old as civilization. As we explore, let's carry with us this profound sense of shared history and the understanding that our bond with the ocean is truly ancient. We're not just exploring a body of water; we're diving into a symbol that has been revered and mythologized by our ancestors, a symbol that connects us all as residents of this blue planet.

The Indo-European Connection

After immersing ourselves in the mythology of ancient Greece, we transition to an exploration of the broader Indo-European connections to the term 'ocean.' The Indo-European languages, spanning multiple continents and cultures, share a common linguistic root. This shared heritage is reflected in language and myths, legends, and perceptions about the natural world, including the ocean.

The dragon, also a common symbol across Indo-European cultures, often represents the forces of nature, and in this context, it is used to depict the encapsulation and control of water bodies. This points to understanding the ocean (and large water bodies) as entities of immense power that nurture and constrain life. They provide resources but also set boundaries, much like the dragon that hoards treasures or controls rivers.

The fascinating connection between Greek 'Ōkeanós' and Vedic' ā-śáyāna-' opens up a wider discussion about how different cultures perceive and represent the ocean. Though geographically separated, these shared linguistic and mythological motifs reveal a collective human understanding of the ocean's power and importance. These common threads weave together a tapestry that reflects our shared human journey with the sea, a testament to the enduring bond between humanity and our planet's watery depths.

In uncovering these linguistic and mythological connections, we begin to appreciate the importance of our relationship with the ocean and its importance to our shared human story. This shared heritage invites us to view the ocean not merely as a physical entity but as a cultural and psychological symbol that continues to shape our worldview.

Chapter 3:
Origin of Water and Formation of the Ocean

Theories on the Origin of Water on Earth

As we look deeper into the heart of our planet's history, we uncover the captivating mystery of how water, the lifeblood of our world, came into existence. Theories abound, each proposing fascinating perspectives and drawing upon diverse fields of study, from astronomy to geology.

There is consensus among scientists that a significant quantity of water would have been present in the material that formed Earth. Still, the exact processes that led to the abundant oceans we see today are subjects of ongoing investigation and debate. As we embark on this exploration, we'll dive into the concept of atmospheric escape, which postulates that water molecules would have escaped Earth's gravity more easily when our planet was less massive during its formation.

We'll also consider the intriguing possibility that Earth may have had magma oceans during its early days of planetary formation. An early atmosphere of carbon dioxide, nitrogen, and water vapor is believed to have formed through the processes of outgassing, volcanic activity, and meteorite impacts. Over millions of years, as Earth's surface cooled, this water vapor would have condensed to form our planet's first oceans.

Picture an Earth where the oceans might have been significantly hotter than today and perhaps even green due to high iron content. Such conjectures provoke our imagination and underscore the elaborate and dynamic history of water on Earth. This journey into the past also offers us valuable insights into the potential existence and form of water on other planets, further expanding the horizons of our understanding.

The Process of Earth's Formation and the Role of Water

A dive into the process of Earth's formation will allow us to uncover the vital role that water has played in shaping our world from its infancy. Earth's formation was a fiery, turbulent process characterized by a massive, rotating disc of dust and gas that slowly coalesced to form our planet. As part of this process, Earth underwent a phase covered in a sea of molten rock.

Water's role might seem elusive amid this fiery world, but it was integral. Scientists theorize that water may have been present in the rocks and minerals that formed Earth, locked away in their crystalline structures. As the Earth cooled and solidified, these water molecules were released, eventually creating the oceans we see today.

Interestingly, it's also believed that a significant portion of Earth's water came from outer space. As our young planet was battered by a steady barrage of comets and meteorites, these cosmic bodies—many of which contained ice—delivered substantial water to our world.

As we explore these theories, we'll discuss the ongoing debates and the scientific evidence that supports each viewpoint. We'll also discuss the fascinating geological evidence that helps scientists constrain the time frame for liquid water on Earth.

This chapter provides an enlightening look at the origins of our oceans and a broader understanding of how our planet was formed, emphasizing the vital and ever-present role that water has played in the story of Earth. Understanding this intricate history is crucial in appreciating the complex interplay of elements that have led to the Earth as we know it today.

The Early Atmosphere of Earth and the Formation of the First Oceans: An Evolutionary Perspective

Earth, our remarkable home in the vast expanse of the cosmos, has witnessed an extraordinary journey from its fiery birth to the world teeming with the life we know today. Central to this journey has been the evolution of water and the formation of oceans. This chapter aims to illuminate the role of geological and atmospheric processes in shaping the Earth's hydrosphere.

A Prelude to Formation

The role of volcanic activity and meteorite impacts in the genesis of Earth's early atmosphere cannot be overstated. These cataclysmic events released a cocktail of gases into the young Earth's atmosphere, primarily composed of carbon dioxide, nitrogen, and water vapor.

The Cooling of Earth and the Birth of the First Oceans: As Earth transitioned from a turbulent world of molten rock to a cooler, more stable entity, conditions began to favor the accumulation of liquid water. The transformation was far from overnight - we are talking about a process that unfolded over millions of years. As the Earth's surface cooled, water vapor trapped in the atmosphere began to condense, and rain started to fall. Over millennia, these rains filled the Earth's depressions, giving birth to the first oceans.

An Alien Seascape: An Imaginative Dive into the Past: One might picture the early oceans as vast, serene bodies of blue just as we see them today, but the reality might have been quite different. Imagine instead seas of a greenish tint, significantly hotter than those we're accustomed to. The reason? High iron content in the waters and a substantially warmer global climate.

Understanding how past oceans differed from today's oceans takes us one step closer to fully appreciating our planet's dynamic and ever-evolving nature. The Journey of Water: From Then to Now: As we trace the journey of water through Earth's history, we come to appreciate its integral role in the evolution of life and the shaping of our planet. From the early rains that filled depressions on Earth's surface to form the first oceans to the ice ages that altered sea levels and shaped our current coastlines, water has been a central character in the life narrative on Earth.

This exploration of the early Earth's atmosphere and the formation of the first oceans is more than just a voyage into the past; it's a journey of understanding the world as it is today. By understanding the genesis of our oceans, we can better appreciate the beauty and complexity of the current marine ecosystems and more accurately predict the future of our planet's watery realms.

Geological Evidence for the Existence of Water on Earth

The existence of water on Earth has been a topic of interest and rigorous investigation by scientists. Unraveling the Earth's past requires detective work of a different kind, with rocks and minerals providing the clues necessary to piece together the history of water on our planet.

The Ancient Witnesses

Rocks - Basalt pillows and the Isua Greenstone Belt: Some of the oldest geological formations on Earth, like basalt pillows and the Isua Greenstone Belt, bear evidence of water's long history on our planet. These structures were formed underwater during volcanic eruptions, providing indisputable evidence that liquid water existed on Earth's surface 3.8 billion years ago.

The Nuvvuagittuq Greenstone Belt

Providing a Glimpse into the Primeval Earth: This geological formation, located in Quebec, Canada, is another window into the early Earth. Rocks here have been dated at an astonishing 4.28 billion years old, reinforcing the theory of water's longstanding presence on our planet.

The Limitations and Challenges of Geological Evidence

Despite the invaluable insights that these geological formations provide, their interpretation is challenging. Earth's dynamic nature means that much of the geological record may have been erased or altered by plate tectonics and erosion. Yet, our evidence suggests that our oceans may have existed in some form since the planet's formation.

This section aims to underscore the significance of geological evidence in determining the history of water on Earth. These ancient stones whisper tales of our world's infancy, and the journey water has embarked on since those early epochs. Through their testimony, we understand more deeply our planet's history and the origin of our life-sustaining oceans.

Chapter 4:
The Ocean's Role in Planetary Formation

The Impact of the Ocean on the Hadean Eon

Navigating the depths of time, we embark on a mesmerizing journey into the Hadean eon, a tumultuous period that witnessed the fiery birth of our planet. Earth was a seething cauldron of volcanic activity and relentless meteorite bombardment during this era. Yet, amidst this chaos, the foundations of our oceans were laid. We will explore how the cooling of the Earth's surface allowed water to condense and gather, eventually giving rise to the vast primordial oceans that shaped the course of our planet's evolution. From the Hadean eon emerged the primordial cradle of life. In this mesmerizing era, water played a defining role in shaping the destiny of our world.

How Oceans Might Have Influenced the Emergence of Life

In the early stages of Earth's history, the oceans held the key to the origin and development of life. These vast bodies of water acted as nurturing crucibles, providing an ideal environment for the emergence of the building blocks of life. We will explore how the oceans might have served as fertile breeding grounds for forming complex organic molecules and the evolution of life forms. From the ancient "primordial soup" to the earliest microorganisms that thrived in these watery atmospheres, we will discover how the oceans played a pivotal role in catalyzing the miracle of life on Earth.

Continuous Change and Adaptation

Casting our gaze upon today's oceans, we come to appreciate their dynamic and ever-changing nature. The ebb and flow of tides, the ceaseless movement of currents, and the interplay of wind and wave paint a vivid portrait of the oceans as living entities intertwined with the Earth's systems. In this section, we will embark on a mesmerizing expedition of the dance of oceanic movements. From the mighty global currents that circulate heat and nutrients across the planet to the mesmerizing beauty of localized phenomena such as upwellings and thermohaline circulation, we will uncover the inner workings of this vibrant aquatic world. We will also delve into the impact of plate tectonics, which, over millennia, have sculpted the ocean basins, shaped the continents, and influenced the configuration of our planet's watery domains.

Additionally, The world ocean, a vast interconnected network of water that spans the globe, is in a perpetual state of transformation due to the relentless forces of geology. Coastlines, the delicate meeting points between land and sea, constantly change as tectonic plates shift, mountains rise and erode, and sea levels fluctuate. Erosion by waves, currents, and wind carves away at rocky cliffs and reshapes sandy shores, while sediment deposition builds new landforms. Subduction zones, where one tectonic plate dives beneath another, give rise to volcanic activity and the creation of islands and island chains. The world ocean's structure is also influenced by seafloor spreading. New crust forms at mid-ocean ridges, expanding the ocean basin over time. These ongoing geological processes leave an indelible mark on the coastline and structure of the world ocean, reminding us of the dynamic nature of our planet and the ever-changing interplay between land and sea.

Past, Present, and Future Oceans

Looking back across the vast tapestry of Earth's history, we witness the mesmerizing story of our oceans. From the ancient Panthalassa, the vast ocean that once surrounded the supercontinent Pangaea, to the familiar configuration of the Atlantic, Pacific, Indian, Southern, and Arctic Oceans that grace our modern maps, the oceans have undergone a remarkable transformation journey. This section will guide us through the evolutionary timeline of our oceans, tracing their changing boundaries, their shifting currents, and the impact they have had on shaping the Earth's climate. Moreover, we will peer into the future, contemplating the challenges ahead. Rising sea levels, ocean acidification, and the implications of climate change pose formidable threats to the delicate balance of our oceans. By understanding the past and present dynamics, we equip ourselves with the knowledge needed to navigate the future and strive for the sustainable stewardship of these precious nautical landscapes.

The Ocean's Unending Story

In concluding this chapter, we find ourselves humbled by the significance of the oceans in the grand narrative of our planet's existence. From their enigmatic origins in the Hadean eon to their indispensable role in nurturing life and shaping the Earth's climate, the oceans stand as resolute custodians of our planet's destiny. As we reflect on the mesmerizing journey through the ocean's role in planetary formation, we are filled with awe and wonder for these vast and majestic bodies of water. Their timeless allure beckons us to embrace our collective responsibility to preserve and protect them, ensuring that future generations may continue to be inspired by their beauty, enriched by their resources, and humbled by their importance in sustaining life on Earth.

Chapter 5:
The Ocean in the Climate System

The Role of the Ocean in Colder and Warmer Periods

Exploring the relationship between the ocean and the climate system, we embark on a journey of how the sea greatly influences Earth's climate during both colder and warmer periods. Throughout Earth's history, the ocean has acted as a crucial regulator, absorbing and storing vast amounts of heat and serving as a primary driver of climate patterns. We will journey through time, examining the ocean's pivotal role in shaping climatic conditions during ice ages, when massive ice sheets covered the land and sea levels were lower. We will also investigate how the ocean responds to and influences climate dynamics during warmer periods, such as interglacial periods, when ice sheets retreat, sea levels rise, and the Earth experiences heightened temperatures. Understanding the ocean's role in these climatic shifts is essential for unraveling the mysteries of the past and comprehending the potential implications of current and future climate change scenarios.

Next, we will explore how the ocean's currents and circulation patterns are pivotal in redistributing heat around the planet. The mighty ocean currents, such as the Gulf Stream and the Kuroshio Current, act as conveyor belts, transporting warm and cold waters across vast distances. These currents influence regional climate patterns and have far-reaching effects on global climate dynamics. We will discover the mechanisms behind these currents, examining the role of wind patterns, temperature variations, and salinity changes in driving their motion. By understanding the complex dance of oceanic circulation, we gain a deeper appreciation for the ocean's vital role in modulating Earth's climate, regulating temperatures, and shaping weather patterns.

Ocean as a Carbon Sink

Beyond its role in heat distribution, the ocean is a crucial player in the global carbon cycle. The ocean absorbs and stores massive amounts of carbon dioxide through various physical and biological processes, effectively acting as a carbon sink. We will investigate how the ocean's capacity to sequester carbon helps to mitigate the impacts of greenhouse gas emissions, buffering the effects of climate change to some extent. However, we will also research the potential consequences of this process, such as ocean acidification, as increased carbon dioxide levels alter the delicate balance of marine ecosystems.

Feedback Mechanisms and Climate Change

The complex relationship between the ocean and the climate system involves feedback mechanisms that can either amplify or dampen climate change. We examine the various feedback loops at play, such as the ice-albedo feedback and the influence of ocean temperatures on atmospheric circulation. By comprehending these feedback mechanisms, we can better understand the potential consequences of anthropogenic climate change on the ocean's dynamics and the cascading effects on global climate patterns.

Additionally, we reflect on the critical role of the ocean in building climate resilience. Understanding the sophisticated interactions between the sea and the climate system is essential for effective climate modeling, prediction, and adaptation strategies. By recognizing the ocean's resilience as a natural climate regulator and protector, we gain insights into how we can work in harmony with this vast aquatic realm to mitigate the impacts of climate change and foster a more sustainable future for both the planet and ourselves.

Effects of Glaciers and Ice Caps on Ocean Levels

Glaciers and ice caps, majestic icy giants that adorn our planet's polar regions and high mountain ranges, significantly influence global ocean levels. In this section, we will probe into the dynamics between these frozen reservoirs and the world's oceans, exploring their role in shaping sea levels and the delicate equilibrium of our planet's climate system.

Glaciers and ice caps are immense accumulations of ice formed from compacted snowfall over thousands of years. They represent vast stores of freshwater that have accumulated on land rather than directly in the oceans. We will discuss how these frozen reservoirs act as natural regulators, storing freshwater during colder periods and releasing it during warmer periods.

As the Earth's climate warms, glaciers and ice caps melt at an accelerated pace. This process contributes to the gradual rise in global sea levels. We also discuss the various mechanisms by which melting ice from glaciers and ice caps enters the ocean, including surface meltwater runoff, calving of icebergs, and the collapse of ice shelves. Through these processes, once stored on land, freshwater finds its way into the world's oceans, resulting in a measurable increase in sea levels.

The effects of melting glaciers and ice caps on ocean levels are not uniform globally. We examine how sea-level changes caused by melting ice can vary regionally, influenced by geographic location, ice thickness, and local climate conditions. Furthermore, we will explore the broader implications of rising sea levels on coastal communities, vulnerable ecosystems, and global climate patterns.

The melting of glaciers and ice caps is linked to climate change through feedback loops. As the ice melts and exposes darker surfaces, such as rock or ocean water, the lower albedo leads to increased absorption of solar radiation, further enhancing the warming effect. We will dig into these feedback processes, studying their role in expediting ice melt and the ensuing elevation of sea levels.

While the effects of melting glaciers and ice caps on ocean levels are significant, they also have far-reaching consequences beyond sea level rise. The freshwater released into the oceans can impact ocean currents, salinity levels, and the delicate balance of marine ecosystems. We will investigate these implications and the potential domino effects on global climate patterns.

By understanding the complex relationship between melting glaciers, ice caps, and ocean levels, we gain insights into the interplay between Earth's cryosphere and hydrosphere. Recognizing the importance of these frozen reservoirs in shaping the delicate equilibrium of our planet's climate system, we can strive for sustainable practices and collective efforts to mitigate the impacts of climate change and protect the future of our oceans and coastal communities.

The Influence of the Ocean on Earth's Atmosphere and Climate

The vast expanse of the ocean not only shapes the Earth's surface but also influences the atmosphere and climate. In this section, we will explore the relationship between the ocean and the Earth's atmospheric conditions, unveiling the ocean's key roles in regulating weather patterns, atmospheric circulation, and the overall stability of our planet's climate system.

The ocean is a massive heat reservoir, storing and releasing vast amounts of thermal energy. Warm ocean currents carry heat from the equator to higher latitudes, moderating temperatures and shaping weather systems. Conversely, cooler ocean currents transport cold water, impacting regional climates and forming distinct climate zones.

Water Vapor and the Water Cycle

The ocean is a primary driver of the Earth's water cycle, which plays a vital role in regulating the distribution of water vapor in the atmosphere. Water is constantly cycled between the ocean, the atmosphere, and land through evaporation, transpiration, and condensation. We will explore how the ocean's vast surface area and water vapor capacity influence cloud formation, precipitation patterns, and the overall humidity of the atmosphere.

Ocean-Atmosphere Interactions

The interface between the ocean and the atmosphere is a dynamic zone where energy, gases, and particles exchange. We will delve into the mechanisms behind these interactions, such as the transfer of gases like oxygen and carbon dioxide and the sale of aerosols and nutrients. These interactions influence air quality, atmospheric composition, and the overall balance of greenhouse gases, playing a crucial role in shaping Earth's climate.

El Niño and La Niña

The tropical Pacific Ocean plays a pivotal role in driving global climate variability through phenomena like El Niño and La Niña. These climate oscillations, influenced by oceanic and atmospheric interactions, have far-reaching impacts on weather patterns, temperature anomalies, and precipitation worldwide.

By unraveling the influence of the ocean on Earth's atmosphere and climate, we gain a deeper appreciation for the interconnectedness of our planet's systems. Understanding the ocean's role as a climate regulator, water vapor distributor, and carbon sink allows us to comprehend the delicate balance that sustains our climate and weather patterns. Through this knowledge, we can better understand the potential consequences of human activities on the ocean and strive for sustainable practices that protect the health and resilience of our oceans and the atmosphere.

Chapter 6:
Geography of the Oceans

A View of Earth as a 'Water World' or 'Ocean World'

Embarking on a captivating exploration of the Earth's watery territories, we explore the awe-inspiring geography of the oceans. As we gaze upon our planet from a cosmic perspective, we are struck by its remarkable characteristic as a "water world" or "ocean world." In this section, we immerse ourselves in the breathtaking realization that more than 70% of Earth's surface is covered by the vast expanse of the oceans. We contemplate the profound implications of this abundant presence of water, acknowledging the oceans as the dominant feature that defines the very essence of our planet.

Boundaries and Divisions

Navigating the expansive blue horizons, we encounter the diverse and distinct divisions that characterize the world ocean. We dig into the geographical delineation of the major oceanic basins, including the Atlantic, Pacific, Indian, Southern, and Arctic Oceans. These vast bodies of water have unique characteristics, currents, and ecosystems, shaping the lives and experiences of those who dwell along their shores.

Seas, Marginal Seas, and Inland Seas

Beyond the expansive oceanic basins, we encounter a rich tapestry of smaller bodies of seawater. These include seas, marginal seas, and inland seas, which add a remarkable diversity to the Earth's watery landscape. We examine the characteristics differentiating seas from oceans, appreciating their partially or fully enclosed nature, often bordered by land. We uncover the unique features of famous seas such as the Mediterranean, Red Sea, and Baltic Sea and their vital contributions to regional ecosystems, maritime trade, and cultural heritage. Moreover, we delve into the fascinating world of marginal seas, which lie at the edges of continents, and inland seas, partially enclosed by landmasses, unfolding the captivating stories they tell about Earth's geological and geographical evolution.

The Continental Shelf - Where the Ocean Meets Dry Land

A significant feature of the ocean's geography is the continental shelf, the transition zone where the vast expanse of the sea meets the solid ground. We study the geological and geographical characteristics of the continental shelf, appreciating its relative shallowness compared to the open ocean. We examine the importance of this region as a habitat for diverse marine life, a source of vital resources, and a critical zone for human activities such as fishing, oil exploration, and underwater research. However, we also confront the challenges and vulnerabilities the continental shelf ecosystems face, as human activities and environmental pressures impact their delicate balance.

Exploring the Depths - The Mysterious Oceanic Abyss

Now we discover the depths of the oceanic abyss, the enigmatic realm that stretches beyond the reach of sunlight. We contemplate the mesmerizing mysteries of the mesopelagic and aphotic zones, where darkness and cold prevail and life adapts to extreme conditions. We marvel at the remarkable creatures that thrive in these deep-sea environments, illuminating our understanding of the incredible biodiversity beneath the surface. Exploring the depths of the oceans, we deeply appreciate their vastness, diversity, and the intrinsic connection between their geography and the web of life they sustain.

As we journey across the oceanic landscapes, we're left in awe by these vast aquatic territories' stunning splendor and deep-seated importance. They shape our planet's surface, influence its climate, and provide a home to an astonishing array of life.

Overview of the Coverage of the Oceans on Earth

The vast and majestic oceans blanket our planet with their awe-inspiring expanse. In this section, we embark on an enlightening overview of the coverage of the oceans on Earth, delving into their immensity and the significance they hold for our planet and all life it supports.

The Dominance of the Oceans

As we contemplate the Earth's surface, we are captivated by the overwhelming dominance of the oceans. Oceans encompass more than 70% of the planet's surface and shape our world's fabric. Their vastness stretches across continents, from expansive coastlines to remote and unexplored regions, forging an interconnected network of water that unites nations, cultures, and ecosystems.

The Five Oceanic Basins

Our exploration leads us to the five major oceanic basins of the world ocean: the Atlantic, Pacific, Indian, Southern, and Arctic Oceans. Each basin possesses unique characteristics, vastness, and marine life, sculpted by the interplay of currents, tides, and geological forces. From the Atlantic's grandeur to the immensity of the Pacific, from the remote Southern Ocean to the frozen Arctic waters, we uncover each oceanic basin's remarkable diversity and distinctiveness.

The Interconnectedness of the Oceans

Beyond their individual distinctions, the oceans are interconnected, forming a global system that transcends boundaries. Powerful oceanic currents, such as the Gulf Stream and the Antarctic Circumpolar Current, circulate vast volumes of water, redistributing heat, nutrients, and marine life on a global scale. We inspect the intense interconnectedness of the oceans, recognizing that what happens in one oceanic basin can have far-reaching effects on others, influencing climate patterns, ecosystems, and the overall health of our planet.

Remote and Unexplored Realms

Despite remarkable advancements in ocean exploration, vast expanses of the oceans remain remote and unexplored. The depths of the abyssal plains, the underwater mountain ranges, and the mysterious trenches that plunge to the deepest points on Earth continue to captivate our curiosity. We reflect on the unknown domains beneath the surface, recognizing the vast potential for discovery and the importance of continued exploration to unravel the mysteries in these untouched corners of the oceans.

Explanation of Point Nemo: The Furthest Point of Inaccessibility in the Ocean

In our exploration of the geography of the oceans, we encounter a fascinating and remote location that holds a unique distinction within the vast expanse of the sea: Point Nemo. Point Nemo is renowned as the furthest point of inaccessibility in the South Pacific Ocean, a place of remarkable solitude and seclusion from human presence.

Geographic Location

Point Nemo, also known as the Oceanic Pole of Inaccessibility, is located at coordinates 48°52.6'S latitude and 123°23.6'W longitude. Its precise location lies approximately 2,688 kilometers (1,670 miles) equidistant from the nearest lands: Ducie Island (part of the Pitcairn Islands), Motu Nui (part of the Easter Island group), and Maher Island (part of Antarctica). It is nestled within the depths of the South

Pacific Gyre, a region characterized by vast expanses of open ocean and minimal human activity.

The Significance of Isolation

Point Nemo's isolation results from its great distance from any landmass, making it remarkably lonely and secluded. Its unique geographic location starkly contrasts the bustling coastlines and populated islands that dot other regions of the world's oceans. Point Nemo's isolation has earned it the reputation of being the most remote location on Earth, far removed from the reaches of civilization and human presence.

Scientific and Environmental Importance

Point Nemo's remote location and isolation make it an ideal site for various scientific and environmental endeavors. Its distance from land minimizes the influence of human activity, providing an unpolluted environment that can serve as a reference point for studying the impact of human presence on other parts of the ocean. Furthermore, the region's deep waters and unique ecological conditions make it an intriguing area for marine research, allowing scientists to investigate marine biodiversity, oceanic processes, and the impacts of climate change.

The Spacecraft Cemetery

Point Nemo has gained additional recognition as the final resting place for decommissioned spacecraft. Due to its remote location and minimal risk of human encounters, space agencies deliberately target this area for the controlled re-entry and disposal of satellites, space stations, and other space debris. Here, remnants of human exploration and technological advancements find their eternal resting place in the ocean's depths, far from inhabited areas and the risk of collision with active satellites.

Reflections on Point Nemo

Point Nemo serves as a reminder of the vastness and solitude within the world's oceans. It captivates our imagination, evoking a sense of wonder and curiosity about the mysteries beneath the surface. This remote and inaccessible location encourages contemplation of the relationship between humanity and the vastness of the natural world. It reminds us of our responsibility to preserve and protect the fragile ecosystems that thrive in the oceans, even in the most remote corners, such as Point Nemo.

Chapter 7:
Dive into the Ocean's Layers

The Division of the Ocean into Vertical and Horizontal Zones

As we journey into the heart of the ocean's enigmatic layers, we uncover an astounding framework that aids our comprehension of its extensive ecosystems and ecological processes: the division of the ocean into vertical and horizontal zones. This intriguing concept illuminates the intricate habitats in the marine world, offering insights into life's distribution, the varying physical conditions, and the dynamic interplay between different oceanic regions.

Oceanographers have distinguished unique zones based on varying physical and biological properties by examining the ocean's vertical dimension. The pelagic zone, a vast, open expanse that extends from the surface to the ocean floor, forms the backbone of the ocean's water column. Within this vast expanse, further sub-divisions emerge, each defined by depth and light penetration, giving birth to realms such as the photic and aphotic zones.

Commencing at the ocean's surface, the photic zone, where light intensity reaches just 1% of surface value, allows for photosynthesis. Plants and microscopic algae convert light, water, carbon dioxide, and nutrients into organic matter. Forming the primary food source for various organisms and playing a vital role in supporting the ocean's rich biodiversity, this region teems with life.

Venturing deeper beyond the photic zone, we encounter the mesopelagic and aphotic zones. The mesopelagic also called the twilight zone, is characterized by a gradual reduction in light levels, yet houses a diverse array of marine life uniquely adapted to low-light conditions. Descending further into the aphotic zone, we find ourselves enveloped in darkness. This mysterious realm is home to extraordinary lifeforms despite its chilling temperatures and stark darkness. These organisms evolved to thrive under extreme pressure and scarcity of resources and embody the resilience of life within the ocean's depth.

Meanwhile, the horizontal dimension offers us a different perspective of the ocean. Here, the ocean is divided into major bodies of water or seas, each enclosed partially or fully by land. The Mediterranean, Caribbean, and Red Seas have unique geographical features, currents, and ecosystems, each painting a distinct picture of marine life. Adding further to this rich tapestry are the marginal and inland seas lying along continental edges and partially enclosed by landmasses.

Exploring these vertical and horizontal zones helps us understand the multifaceted yet beautifully interconnected marine environments. They provide a robust framework for conserving the ocean's ecosystems while unraveling the secrets of adaptation and survival in various oceanic regions. Every zone provides a window into the diverse and complex marine life, offering a deeper appreciation of the resilience and marvel of organisms thriving in the ocean's vast aquatic realm. As we peel back the layers of the ocean, we're reminded of the enchanting mysteries that remain hidden within each zone.

Chapter 8:
Oceanic Temperatures and Currents

Factors Affecting Ocean Temperatures

As we delve into the fascinating world of oceanic temperatures, we encounter many factors that influence the heat distribution within the vast ocean. Understanding these factors allows us to comprehend the complex dynamics of temperature variations and their impact on marine life, climate patterns, and global oceanic circulation.

Solar Radiation

Solar radiation plays a fundamental role in determining ocean temperatures. The amount of sunlight absorbed by the ocean's surface directly affects its heating. Near the equator, where sunlight is more direct, the ocean experiences higher temperatures. As we move towards the poles, the sun's angle decreases, resulting in lower temperatures. Variations in solar radiation throughout the year, influenced by the Earth's axial tilt and seasonal changes, contribute to the cyclical nature of oceanic temperature patterns.

Latitude and Geographical Position

Latitude and geographical position have a profound impact on ocean temperatures. Regions closer to the equator, characterized by lower margins, tend to experience warmer temperatures due to their proximity to the sun's direct rays. In contrast, higher latitudes near the poles receive less solar radiation, resulting in cooler temperatures. The oceanic temperature gradient across spaces contributes to the formation of global climate patterns and drives the circulation of ocean currents.

Oceanic Currents

Ocean currents, driven by various forces such as wind, temperature differences, and the Earth's rotation, significantly influence ocean temperatures. Currents act as conveyer belts, transporting warm or cold water from one region to another. Warm ocean currents, such as the Gulf Stream, carry heat from the equatorial regions to higher latitudes, raising the temperatures of coastal areas. Conversely, cold ocean currents, like the California Current, bring cooler waters from the polar regions, leading to lower temperatures along their paths.

Land-Sea Interactions

The proximity of landmasses to the ocean can influence local temperature patterns. Coastal areas are subject to the moderating effect of the ocean, which can regulate temperature extremes. Land-sea breezes, generated by differential heating between land and water, can impact coastal temperatures. During the day, the land heats faster than the ocean, leading to onshore breezes and cooler temperatures. At night, the reverse occurs, resulting in offshore breezes and relatively warmer temperatures.

By comprehending the factors contributing to oceanic temperature variations, we gain insights into the complex interplay between the sun, currents, and geographical features. These temperature dynamics shape climate patterns, influence marine habitats, and drive the mechanisms of the Earth's hydrosphere. Exploring the depths of oceanic temperatures unveils the delicate balance and interconnectedness of the world's oceans, inspiring us to appreciate the large impact these factors have on the fragile equilibrium of our planet.

Explanation of Major Ocean Currents and Their Roles in Global Climate

Ocean currents, powerful and dynamic, play a crucial role in shaping the Earth's climate system. In this section, we dive into the depths of major ocean currents, exploring their pathways and understanding their significant contributions to global climate patterns.

The Gulf Stream

One of the most prominent and well-known ocean currents, the Gulf Stream, flows along the eastern coast of North America before turning northeastward toward Europe. Originating in the Gulf of Mexico, this warm and swift current transports vast amounts of heat from the tropics to higher latitudes, influencing the climate of both regions. The Gulf Stream's warm waters contribute to the temperate climate of Western Europe, making it milder than other areas at similar latitudes.

The Kuroshio Current

The Kuroshio Current, flowing northward along the eastern coast of Asia, is Japan's equivalent of the Gulf Stream. Originating in the Philippine Sea, this warm, fast-moving current carries heat from the tropics to the western North Pacific Ocean. Its influence extends beyond temperature regulation, shaping the region's climate and supporting diverse marine ecosystems.

The Agulhas Current

The Agulhas Current, located along the eastern coast of South Africa, is a warm and powerful western boundary current. Originating from the Indian Ocean, it transports warm water southward, parallel to the coast. The Agulhas Current influences regional climate, providing heat and moisture for nearby land areas. It also plays a role in the global thermohaline circulation, contributing to the movement of warm surface waters toward the poles.

The Antarctic Circumpolar Current

The Antarctic Circumpolar Current, encircling Antarctica, is the world's largest ocean current. This cold and swift eastward-flowing current connects the Atlantic, Indian, and Pacific Oceans, serving as a vital link between the major oceanic basins. The Antarctic Circumpolar Current influences global climate by transferring large amounts of cold water from the Southern Ocean to lower latitudes, affecting temperature distributions and regulating atmospheric carbon dioxide levels.

Oceanic Upwelling and Downwelling

Upwelling and downwelling phenomena impact oceanic temperatures, particularly in coastal areas. Upwelling occurs when cold, nutrient-rich waters rise to the surface, replacing warmer surface waters. This process promotes cooler temperatures and enhances marine productivity by bringing nutrient-laden waters from the depths. Downwelling, on the other hand, involves sinking surface waters, often associated with the convergence of currents or the formation of deep-ocean circulation patterns. Downwelling can lead to the transfer of heat from the surface to deeper layers, affecting temperature distributions.

The interactions of these major ocean currents and upwelling and downwelling processes contribute to the complex and interconnected system known as the global thermohaline circulation. This system is vital in distributing heat around the planet, influencing regional climates, and impacting weather patterns. The ocean's immense heat capacity and the movement of oceanic currents act as a giant heat reservoir, helping to moderate temperature extremes, regulate atmospheric conditions, and shape the Earth's climate locally and globally.

As we demystify the sophisticated web of major ocean currents and their significant impact on global climate, we fully appreciate the complex interplay between the oceans, atmosphere, and land. These currents serve as the essential engines behind climate variability, sculpting weather systems and dictating the transfer of heat and moisture across the globe. Grasping this delicate equilibrium and the interdependencies among these currents prompts us to acknowledge the oceans' influence on our climate. It further underscores the urgent necessity to protect and manage these water bodies sustainably, ensuring the health of our planet and the well-being of generations to come.

Chapter 9:
Gas Exchange in the Ocean

Overview of the Dissolved Gases in the Ocean

As we venture into the ocean's depths, we discover a hidden world where gases dissolve in the watery embrace. The ocean, a vast reservoir of dissolved gases, plays a crucial role in global biogeochemical cycles and the delicate balance of Earth's atmosphere. This chapter explores the fascinating array of gases that find solace within the ocean's depths.

Oxygen

Oxygen is one of the most vital gases present in the ocean. Through photosynthesis, marine plants and algae release oxygen into the water, allowing it to dissolve and support the respiration of marine organisms. Dissolved oxygen is essential for the survival of marine life, providing the energy required for metabolic processes and maintaining the delicate balance of ecosystems.

Carbon Dioxide

Carbon dioxide, a greenhouse gas, is also present in the ocean in dissolved form. Carbon dioxide enters the ocean through various processes, including respiration and the dissolution of atmospheric CO_2. The ocean serves as a significant sink for carbon dioxide, helping to regulate its concentration in the atmosphere and mitigate the impacts of climate change. However, increasing carbon dioxide levels due to human activities have led to ocean acidification, altering the delicate pH balance and posing challenges for marine organisms and ecosystems.

Nitrogen

Nitrogen is another crucial gas found in the ocean. It enters the water through various sources, including atmospheric deposition, rivers, and nitrogen-fixing organisms. Dissolved nitrogen is vital in supporting primary productivity and the growth of marine plants and algae. It is an essential nutrient for aquatic organisms, contributing to marine ecosystems' overall health and balance.

Other Gases

The ocean also contains other dissolved gases, albeit in smaller concentrations. These include methane, hydrogen sulfide, and various trace gases. Certain microorganisms in oxygen-depleted environments, such as wetlands and marine sediments, can produce methane, a potent greenhouse gas. Hydrogen sulfide, with its characteristic pungent odor, can be present in oxygen-deficient regions, such as oxygen minimum zones or near hydrothermal vents.

The concentrations of dissolved gases in the ocean vary spatially and temporally, influenced by temperature, salinity, biological activity, and gas exchange with the atmosphere. The solubility of gases in seawater depends on these factors, with colder water generally having a higher capacity for gas dissolution. Understanding the dynamics of gas exchange in the ocean is crucial for studying the impacts of climate change, assessing marine ecosystem health, and predicting the response of marine organisms to environmental changes.

Venturing into the world of dissolved gases in the ocean allows us to discern the complex interconnections between the marine world, the atmosphere, and the myriad organisms that inhabit the ocean's depths. The ocean's dual function as a reservoir and regulator of dissolved gases underscores its pivotal role within the Earth's climate framework. Exploring the mechanisms and patterns of gas exchange in the ocean enriches our understanding of our planet's functionality and the intricate interplay of its multifaceted ecosystems.

Discussion on Ocean Acidification Due to Increased Carbon Dioxide Concentration

Ocean acidification, a consequence of rising carbon dioxide (CO_2) atmospheric concentrations, poses significant challenges to the delicate balance of marine ecosystems. In this section, we study the phenomenon of ocean acidification, exploring its causes, impacts, and potential implications for marine life.

When excess CO_2 from human activities is absorbed by the ocean, a chemical reaction occurs, decreasing seawater pH. This process alters the ocean's carbonate chemistry and reduces the concentration of carbonate ions. These carbonate ions are crucial for forming calcium carbonate, a building block for marine organisms, including corals, shellfish, and phytoplankton.

The repercussions of ocean acidification are far-reaching. One of the most notable impacts is the degradation of coral reefs, which are highly sensitive to changes in pH levels. As the ocean becomes more acidic, corals struggle to build and maintain their calcium carbonate skeletons, leading to coral bleaching, reduced growth rates, and increased vulnerability to other stressors. The loss of coral reefs has intense implications for the biodiversity and resilience of marine ecosystems, as they provide essential habitats for a vast array of marine species.

Ocean acidification also affects shell-forming organisms, such as oysters, clams, and certain plankton species. These organisms rely on calcium carbonate to form their shells or protective exoskeletons. Decreasing carbonate availability compromises their ability to build and maintain these structures, making them more susceptible to predation, disease, and reduced survival rates. This disruption in the marine food web can have cascading effects on entire ecosystems, impacting fisheries, coastal economies, and the livelihoods of communities dependent on marine resources.

Furthermore, ocean acidification can affect the physiology and behavior of marine organisms. Studies suggest that acidification can impair sensory and metabolic functions, alter growth rates, and reduce reproductive success in various species. These changes in physiological processes can disrupt ecological interactions and compromise the overall health and stability of marine ecosystems.

Addressing the issue of ocean acidification requires concerted efforts to mitigate CO2 emissions and reduce the amount of carbon dioxide entering the atmosphere. Implementing sustainable practices, transitioning to cleaner energy sources, and protecting marine habitats are crucial steps toward safeguarding the health of our oceans. Additionally, research and monitoring programs are vital for understanding the impacts of acidification on marine ecosystems and informing conservation strategies.

By recognizing the profound impact of ocean acidification on marine life, we are compelled to take action to preserve the delicate balance of our oceans. Through collective responsibility and sustainable practices, we can mitigate the risks posed by increasing carbon dioxide concentrations and protect the diverse and vibrant ecosystems that rely on the health of our oceans.

Chapter 10:
The Ocean's Rich Biodiversity

Examination of the Diverse Species Living in the Ocean

The ocean, an awe-inspiring realm of wonder, harbors a staggering array of life forms. In this chapter, we embark on a journey to explore the rich biodiversity that thrives beneath the waves, uncovering the extraordinary diversity of species that call the ocean their home.

The oceanic environment supports many organisms, ranging from the tiniest microorganisms to colossal marine mammals. Phytoplankton, microscopic algae that harness the power of sunlight through photosynthesis, form the foundation of the marine food web. These microscopic organisms, along with their counterparts, zooplankton, nourish a vast array of marine life, including fish, invertebrates, and even larger predators.

Countless species have adapted to various habitats and ecosystems within the ocean's depths. Coral reefs, vibrant and teeming with life, house an astonishing diversity of marine species. These delicate ecosystems support many colorful fish, corals, crustaceans, and other invertebrates. Coral reefs are biodiversity hotspots and serve as natural coastal barriers, protecting shorelines from erosion and storm damage.

The open ocean, or pelagic zone, is a vast expanse where migratory species roam and elusive creatures dwell. Majestic marine mammals, such as whales, dolphins, and seals, traverse great distances, undertaking remarkable migrations across ocean basins. These magnificent creatures captivate our imagination and remind us of the interconnectedness of life in the ocean.

The deep sea, a realm shrouded in darkness and extreme conditions, harbors mysterious and often bizarre organisms. Strange-looking fish, bioluminescent creatures, and unique adaptations abound in this extraordinary habitat. Venturing into the depths reveals astonishing discoveries, such as hydrothermal vent communities, where life thrives without sunlight, relying on chemosynthesis to harness energy from chemical reactions.

The ocean is estimated to be home to over 230,000 identified species, with countless more yet to be discovered. From tiny plankton to immense whales, from delicate corals to resilient deep-sea organisms, the ocean's biodiversity is a testament to the resilience and adaptability of life. Each species plays a vital role in maintaining the health and balance of marine ecosystems, contributing to processes such as nutrient cycling, carbon sequestration, and the overall functioning of the planet.

However, the ocean's rich biodiversity faces numerous threats, including habitat degradation, overfishing, pollution, and the impacts of climate change. Human activities have put immense pressure on marine ecosystems, compromising the delicate balance of life beneath the waves. Through scientific exploration, conservation initiatives, and raising awareness, we can appreciate the awe-inspiring diversity of life in the ocean and work towards preserving it for future generations.

Impact of Human Activity on Ocean Ecosystems, Including Pollution and Overfishing

The delicate equilibrium of ocean ecosystems, with their complex life network, is progressively undermined by human endeavors. In this segment, we explore the significant implications of human-induced activities on the vitality and endurance of ocean ecosystems, spotlighting key issues such as pollution and overfishing.

Pollution

Human-generated pollution has become a significant threat to the oceans and the organisms that depend on them. Industrial and agricultural runoff, coastal development, and improper waste disposal have accumulated harmful substances in marine environments. Chemical pollutants, including heavy metals, pesticides, plastics, and oil spills, contaminate the water and pose serious risks to marine life. These pollutants can disrupt reproductive systems, impair growth and development, and even cause mortality in aquatic organisms. Additionally, plastic debris, which takes centuries to degrade, accumulates in marine ecosystems, endangering marine species through entanglement and ingestion.

Overfishing

Overfishing, driven by commercial demand and unsustainable fishing practices, has severely depleted fish populations worldwide. Large-scale fishing operations, including bottom trawling and longlining, indiscriminately capture vast amounts of marine life, often resulting in the depletion of target species and collateral damage to non-target species. Removing key predator species disrupts the balance of marine food webs, leading to cascading effects throughout the ecosystem. Overfishing also impacts the livelihoods of coastal communities that rely on fish stocks for food security and economic stability.

Habitat Destruction

Human activities, such as bottom trawling, coastal development, and destructive fishing practices, destroy critical marine habitats. Physical damage, sedimentation, and pollution threaten coral reefs, seagrass meadows, mangrove forests, and other essential ecosystems. The loss and degradation of these habitats jeopardize the survival of numerous species that depend on them for food, shelter, and reproduction. The destruction of coastal habitats also leaves shorelines vulnerable to erosion and storm damage, putting human communities at risk.

Climate Change

The ocean is intricately linked to Earth's climate system, and human-induced climate change is altering oceanic conditions at an alarming rate. Rising sea temperatures, ocean acidification, and sea-level rise significantly affect marine ecosystems. Warmer waters disrupt marine species' distribution and behavior, causing range shifts and impacting their reproductive cycles. Ocean acidification threatens the growth and survival of organisms that rely on calcium carbonate structures, such as corals and shellfish. Additionally, sea-level rise poses risks to coastal habitats and communities, increasing the vulnerability of coastal ecosystems and exacerbating the impacts of storms and flooding.

Mitigating the Impacts

Addressing these challenges requires concerted efforts to mitigate pollution, promote sustainable fishing practices, protect critical habitats, and reduce greenhouse gas emissions. Implementing stricter regulations and adopting innovative technologies can help reduce pollution inputs into the oceans. Sustainable fisheries management involves establishing marine protected areas and utilizing ecosystem-based approaches to support the recovery of fish stocks and protect marine biodiversity. Conservation initiatives aimed at restoring and preserving critical habitats are essential for maintaining the resilience and functionality of marine ecosystems. Additionally, reducing greenhouse gas emissions and transitioning to cleaner energy sources can help mitigate the impacts of climate change on the oceans.

By recognizing the detrimental effects of human activities on ocean ecosystems, we can take collective action to promote responsible stewardship of these vital environments. Preserving the health and biodiversity of the oceans is crucial for the survival of marine species and the well-being of human communities that depend on the oceans for food, livelihoods, and cultural identity. Through sustainable practices, awareness campaigns, and international cooperation, we can work towards a future where the oceans thrive, ensuring the preservation of our planet's most valuable ecosystems.

Chapter 11:
Legends of the Deep

Exploration of Various Oceanic Legends and Myths

Throughout history, the vast and enigmatic oceans have sparked the human imagination, giving rise to many captivating legends and myths. In this chapter, we set sail on a voyage to untangle the tales interwoven into the rich fabric of maritime folklore, enveloping us in the sense of wonder and enigma that pervades the depths.

One of the most enduring legends of the sea is that of the Kraken, a colossal sea creature believed to dwell in the deep ocean. Described as a massive, tentacled beast capable of capsizing ships, the Kraken has captured the imaginations of sailors and storytellers for centuries. Legends of encounters with this mythical creature have been passed down through generations, adding an air of mystique to the vast expanses of the ocean.

The fabled Lost City of Atlantis, often depicted as an advanced and prosperous civilization submerged beneath the sea, has fascinated explorers and scholars for ages. The ancient Greek philosopher Plato described Atlantis as a utopian society that met its tragic fate in a cataclysmic event. The story of Atlantis continues to captivate our imagination, prompting quests for its discovery and inspiring countless literary and artistic works.

Tales of sea monsters have been prevalent in maritime folklore across cultures. From the sea serpents of Norse mythology to the mythical Leviathan and the legendary sea dragons of Asian legend, these captivating creatures embody the unknown and untamed aspects of the ocean. They symbolize the power and untapped mysteries that lie beneath the waves.

The enchanting allure of sirens and mermaids has bewitched seafarers throughout history. These mythical creatures, often depicted as half-human and half-fish, embody both beauty and danger. Legends tell of their mesmerizing songs that lured sailors to their doom or their benevolent presence guiding lost souls to safety. The tales of sirens and mermaids blur the lines between fantasy and reality, stirring our fascination with the ocean's secrets.

The legend of the Flying Dutchman, a ghost ship doomed to sail the seas for eternity, has become an enduring symbol of maritime lore. The ghostly vessel said to be crewed by tormented souls, is believed to be an omen of misfortune and doom to those who encounter it. The tale of the Flying Dutchman speaks to the mysteries of the ocean and the timeless stories of lost souls trapped between the realms of the living and the dead.

These oceanic legends and myths reflect our deep-seated fascination with the unknown and our desire to make sense of the sea's vast and often treacherous realm. They serve as a testament to the enduring power of storytelling and the human need to weave narratives that explore the uncharted territories of our imagination. While these legends may originate from fantasy, they evoke a sense of wonder and ignite our curiosity about the wonders beneath the waves.

Immersing ourselves in the enthralling narratives of the deep, we begin to grasp the cultural importance embedded within these legends, transcending boundaries and generations. They serve as poignant reminders of the deep-rooted bond between humanity and the ocean, evoking a range of emotions and instilling an appreciation for the aquatic domain's immeasurable power and enigmatic wonders. These timeless tales of the deep fuel our relentless quest for exploration, igniting our imagination and nurturing an unbreakable connection with the vast and captivating world that is the ocean.

The legends and myths surrounding the ocean have allowed humans to grapple with the mysteries and complexities of this vast and untamed realm. These tales offer a glimpse into the connection between humanity and the enigmatic world beneath the waves. The ocean, with its vastness and hidden depths, has always been a source of wonder and awe. Legends such as the Kraken and sea monsters symbolize the unfathomable creatures that inhabit the depths, which remain largely unexplored and mysterious. These tales reflect our fascination with the unknown and our deep-seated desire to unravel the secrets beneath the surface. They remind us that despite our advancements in science and technology, hidden wonders and creatures still defy our understanding.

The myth of Atlantis, a lost city beneath the sea, speaks to the allure of hidden civilizations and submerged landscapes. It represents the notion that beneath the waves, there may exist ancient ruins and forgotten worlds waiting to be discovered. The legend of Atlantis taps into our collective imagination. It fuels our desire to explore and unravel the mysteries of the ocean's depths.

Sirens and mermaids, with their seductive songs and mythical allure, capture the beauty and danger that coexist in the marine world. These legends highlight the captivating power of the ocean and its ability to mesmerize and enthrall. They evoke a sense of enchantment and mystery, reminding us of the unfathomable wonders that lie beneath the waves.

The tale of the Flying Dutchman and other ghostly apparitions at sea reflects the sense of mystery and the inherent risks associated with maritime exploration. These legends embody the ethereal and often treacherous nature of the ocean, reminding us of the dangers and uncertainties that sailors and explorers have faced throughout history. They serve as a reminder of the vastness and unpredictability of the ocean, leaving room for the unexplained and the supernatural.

As the ocean's mysteries intertwine, the oceanic legends evoke a sense of reverence and respect for the power and complexity of the maritime realms. They inspire exploration and curiosity, urging us to delve deeper into the unknown and uncover the truths beneath the waves. While these tales may be rooted in folklore and mythology, they speak to our deep connection with the ocean and our innate desire to seek answers and make sense of the world.

Conclusion

The Wonder and Responsibility of Oceans

Throughout our exploration of the oceans, we've marveled at their beauty, ecological significance, and significant mysteries. From climate regulation to cultural importance, the oceans are essential to our planet and well-being.

The oceans are our life support system covering over 70% of the Earth's surface. They produce oxygen, regulate temperatures, and absorb carbon dioxide. They provide valuable resources and serve as trade, transportation, and cultural exchange avenues. The oceans are a source of inspiration, recreation, and solace for countless people.

However, our oceans face numerous challenges. Pollution, overfishing, climate change, and habitat destruction threaten their delicate balance. Our responsibility as custodians of the Earth is to protect and preserve these invaluable ecosystems.

To safeguard the oceans, we must adopt sustainable practices. Minimizing pollution, promoting responsible fishing, and reducing our carbon footprint are crucial steps. Conservation and restoration efforts should focus on critical habitats like coral reefs, mangroves, and seagrass meadows. Embracing renewable energy sources and international cooperation in addressing climate change is vital.

Education and awareness are crucial in fostering stewardship and appreciation for the oceans. Understanding our interconnectedness with them inspires action. Citizen science initiatives, marine research support, and community-driven conservation efforts make a difference.

Our responsibility to the oceans extends beyond our lifetime. It's a commitment to future generations and the sustainability of our planet. We can forge a harmonious relationship with the oceans by embracing this responsibility.

Call to Action for Ocean Preservation

The oceans' plight demands immediate action. We call upon individuals, communities, governments, and organizations to safeguard the oceans.

Raise Awareness: Share knowledge, educate, and promote dialogue about the oceans' value and vulnerabilities.

Reduce Pollution: Embrace sustainable practices, recycle, minimize single-use plastics, and support responsible waste management.

Promote Sustainable Fisheries: Choose certified sustainable seafood, advocate for fishing regulations, and support sustainable aquaculture.

Protect Critical Habitats: Recognize the importance of coastal ecosystems, coral reefs, mangroves, and seagrass meadows. Support initiatives and engage in community-led conservation efforts.

Mitigate Climate Change: Support renewable energy, reduce greenhouse gas emissions, and advocate for climate action.

Foster International Collaboration: Encourage global cooperation in marine research, policy development, and conservation efforts.

Support Ocean Research and Conservation: Back organizations dedicated to marine research, contribute to citizen science initiatives, and volunteer or donate to projects focused on ocean protection.

Together, we can make a difference. Let's ensure the oceans thrive, biodiversity flourishes, and their immeasurable value is respected. The time to act is now. Together, we can be the guardians our oceans need.

Resources

Ballard, R.D. (2001). Lost Liners: From the Titanic to the Andrea Doria, the Ocean Floor Reveals Its Greatest Ships. Hyperion.

Clendenning, A. (2021). Twenty fun facts about the ocean. *BYUH Ke Alaka'i.* https://kealakai.byuh.edu/twenty-fun-facts-about-the-ocean

Cousteau, J.Y. (1953). The Silent World: A Story of Undersea Discovery and Adventure. Harper & Brothers.

Earle, S.A. (2010). The World Is Blue: How Our Fate and the Ocean's Are One. National Geographic.

Ellis, R. (1998). The Empty Ocean: Plundering the World's Marine Life. Island Press.

Gessner, D. (2019). The Book of Atlantis Black: The Search for a Sister Gone Missing. Tin House Books.

Hohn, D. (2019). Moby-Duck: The True Story of 28,800 Bath Toys Lost at Sea and of the Beachcombers, Oceanographers, Environmentalists, and Fools, Including the Author, Who Went in Search of Them. Penguin Books.

McPhee, J. (1997). The Founding Fish. Farrar, Straus and Giroux.

Montgomery, S. (2015). The Soul of an Octopus: A Surprising Exploration into the Wonder of Consciousness. Atria Books.

Safina, C. (2016). The View from Lazy Point: A Natural Year in an Unnatural World. Henry Holt and Company.

Team, T. (2023). 10 Unbelievable Facts about the Ocean. *Real Word*. https://www.trafalgar.com/real-word/10-unbelievable-facts-ocean/

Top 10 things you didn't know about the ocean | U.S. Geological Survey. (n.d.). https://www.usgs.gov/programs/cmhrp/news/top-10-things-you-didnt-know-about-ocean

Wikipedia contributors. (2023c). Ocean. *Wikipedia*. https://en.wikipedia.org/wiki/Ocean

Winchester, S. (2010). Atlantic: Great Sea Battles, Heroic Discoveries, Titanic Storms, and a Vast Ocean of a Million Stories. Harper Perennial.

The Abyss Inside:

Mind-Blowing Facts and Discoveries
About Your Extraordinary Human Body

Introduction

Welcome to "The Abyss Inside: Mind-Blowing Facts and Discoveries About Your Extraordinary Human Body." This is an exhilarating journey through the depths of human anatomy and physiology. Get ready to explore the intricate wonders within you as we unravel the mysteries and unveil the unique facts about your miraculous human body.

The Importance of Understanding the Human Body

Have you ever wondered what makes you who you are? With its awe-inspiring complexity, your body is not merely a vessel but a masterpiece of evolution. Understanding how this machinery functions is fascinating and essential for nurturing a healthier, more vibrant life.

By delving into the intricacies of the human body, we gain insights into how it operates, allowing us to make informed choices about our well-being. From unlocking the secrets of disease prevention to optimizing our physical and mental performance, this knowledge empowers us to take charge of our health.

This captivating book will explore the human body comprehensively, examining its various systems, organs, and functions. Each chapter will take you on a thrilling adventure through the network of bones, muscles, nerves, and organs that make up this astonishing vessel we call our body.

From the fascinating structures of the skeletal system to the pulsating rhythm of the circulatory system, from the workings of the nervous system to the complex dance of hormones in the endocrine system, we will uncover mind-blowing facts, discoveries, and intriguing insights that will leave you in awe of your own existence.

Prepare to be amazed as we delve into the depths of human anatomy, discover the mysteries of bodily functions, and unveil the symphony of systems that work harmoniously to sustain your life. Along the way, we will explore the fascinating growth, development, and aging processes, shedding light on the profound changes that occur throughout the human lifespan.

Prepare to buckle up and gear yourself for an enthralling journey into "The Abyss Inside: Mind-Blowing Facts and Discoveries About Your Extraordinary Human Body." Get set to be spellbound, enlightened, and ignited by the breathtaking marvels within you.

Chapter 1:
Human Anatomy

Definition and Significance of Human Anatomy

Beneath the surface of our existence, a complex mosaic of components unfolds, each assigned a unique role and functionality. Step into the fascinating world of human anatomy as we examine the body's detailed architecture and comprehend its deep-seated importance.

Human anatomy is the study of the structure of the human body, examining the organization and arrangement of its various components. Through this exploration, we gain insights into how our bodies are constructed, from the microscopic building blocks to the grand design of interconnected systems.

Understanding human anatomy gives us a deeper appreciation for the incredible precision and complexity that allows our bodies to function seamlessly. It opens a gateway to knowledge about the body's diverse organs, tissues, and cells, unveiling their unique roles and contributions to our well-being.

But beyond its scientific importance, human anatomy holds profound significance. It allows us to connect with our bodies, fostering a sense of self-awareness and understanding. Through this exploration, we can truly marvel at the intricacies of our existence and develop a newfound respect for the vessel that carries us through life.

Overview of the Human Body's Structure and Organization

To fully understand the marvels of the human body, it's essential to comprehend its incredible structure and arrangement. In this portion, we'll set off on an illuminating voyage through the complexities of our physiological structure, revealing the astonishing framework that forms the foundation of our being.

The human body is a masterpiece of design, composed of various systems working harmoniously to sustain life. The encompassing array of systems comprises the skeletal system, muscular system, circulatory system, nervous system, respiratory system, digestive system, reproductive system, and numerous others. Each method has unique organs, tissues, and cells contributing to its specific functions.

The skeletal system is at the core of our physical framework, which provides support, protection, and mobility. This thorough network, composed of bones, cartilage, ligaments, and tendons, forms the foundation upon which our bodies are built. It gives us our shape and serves as a reservoir for essential minerals, produces blood cells, and houses the bone marrow.

The muscular system, collaborating with the skeletal system, allows us to make movements, from the most delicate gestures to the most energetic actions. Muscles, which are composed of fibrous tissues, contract and relax, thus generating the power necessary for various bodily movements. This intricate system, encompassing hundreds of muscles throughout our body, equips us with the capacity to walk, run, grip, and execute many other tasks.

The circulatory system facilitates the distribution of oxygen, nutrients, and waste products. It consists of the heart, blood vessels, and blood working together to ensure the transportation of vital substances throughout the body. The heart acts as a powerful pump, propelling oxygenated blood to tissues and organs while simultaneously receiving deoxygenated blood to be reoxygenated in the lungs.

The nervous system, comprising the brain, spinal cord, and a network of nerves, serves as the body's communication and control center. It enables us to perceive the world, process information, and coordinate our actions. From sensory perception to motor coordination, this system plays a fundamental role in our daily lives.

These are just a few glimpses into the remarkable organization and interplay of the human body's systems. In the upcoming chapters, we will delve deeper into each design, uncovering its functions, connections, and the fascinating discoveries that have expanded our understanding of ourselves.

Get ready to be astounded as we investigate the remarkable structure and arrangement of the human body. Collectively, let's decipher the enigmas beneath our skin and develop a deep admiration for the complex harmony of systems that constitute our identity.

Skeletal System and Its Role in Providing Support and Protection

The skeletal system serves as the foundation of our bodies, providing essential support, structure, and protection. Within its framework of bones lies a remarkable network that gives us our shape, safeguards our delicate internal organs, and facilitates movement.

The skeletal system, composed of bones, cartilage, ligaments, and tendons, is crucial in maintaining our overall physical integrity. Its primary functions extend beyond mere support; it serves as a mineral reservoir, produces blood cells, and facilitates bodily movements.

One of the key functions of the skeletal system is to provide structural support to the body. The bones form a sturdy framework that gives our bodies their shape and posture. They act as a scaffold for muscles, allowing them to attach and generate movements. Without the support of the skeletal system, our bodies would lack the stability required for everyday activities, such as standing, walking, and performing various tasks.

Moreover, the skeletal system serves as a protective shield for our internal organs. It encases vital structures, such as the brain, heart, lungs, and abdominal organs, shielding them from external impact and potential injury. The skull protects our brain, the ribcage safeguards our heart and lungs, and the vertebral column covers the delicate spinal cord.

In addition to support and protection, the skeletal system plays a critical role in locomotion. Bones, along with the muscles and joints, enable movement and provide leverage for muscle contraction. Through the coordinated action of bones and muscles, we can walk, run, jump, and perform detailed activities with precision and control.

Furthermore, the skeletal system contributes to the production of blood cells through a process called hematopoiesis. Within the bone marrow, specialized cells continuously generate new red and white blood cells, vital for oxygen transport, immune function, and overall health.

Muscular System and Its Function in Movement

The human body possesses a phenomenal ability to move and perform various physical activities, from the subtlest gestures to the most dynamic athletic feats. At the core of this remarkable capability lies the muscular system, a complex network of muscles that enables us to bend, flex, extend, and contract with precision and power.

The muscular system consists of three main types of muscles: skeletal, smooth, and cardiac. In this section, we will focus primarily on skeletal muscles responsible for voluntary movement and play a fundamental role in our daily lives.

Skeletal or striated muscles are attached to bones by tendons and work in pairs to produce controlled movements. These muscles are composed of individual muscle fibers that contract and relax in response to nerve signals, resulting in coordinated and purposeful actions.

The primary function of the muscular system is to generate movement and maintain posture. When skeletal muscles contract, they pull on the bones, causing them to move. By working in synergy with the skeletal system, muscles allow us to perform an incredible array of activities, such as walking, running, lifting, and even sophisticated tasks like playing a musical instrument or typing on a keyboard.

Beyond movement, the muscular system also contributes to the stability and posture of our bodies. Certain muscles, known as postural muscles, work continuously to support the alignment of our spine and maintain an upright position against the force of gravity. These muscles are crucial in preventing postural imbalances and ensuring proper body alignment.

Additionally, the muscular system is responsible for generating heat. When muscles contract and relax, they produce heat as a byproduct of their metabolic processes. This heat production helps to regulate body temperature, allowing us to maintain a constant internal environment, even in fluctuating external conditions.

The muscular system also plays a vital role in protecting internal organs. Muscles encompass essential structures, such as the heart, composed of specialized cardiac muscles. The rhythmic contractions of cardiac muscles ensure continuous blood circulation throughout the body, supplying oxygen and nutrients while removing waste products.

Furthermore, the muscular system supports essential bodily functions beyond voluntary movement. Smooth muscles, found in the walls of organs, blood vessels, and the digestive system, enable involuntary movements, such as the rhythmic contractions of the digestive tract and the dilation and constriction of blood vessels.

Nervous System and Its Role in Communication and Coordination

The human body is a complex network of interconnecting systems, and at the helm of this sophisticated web lies the nervous system. With its remarkable ability to transmit signals and coordinate activities, the nervous system serves as the master conductor of the human orchestra.

The nervous system is responsible for communication, integration, and control throughout the body. It consists of two primary components: the central nervous system (CNS), which includes the brain and spinal cord, and the peripheral nervous system (PNS), which encompasses the network of nerves that extend throughout the body.

At the core of the nervous system is the brain, the epicenter of intelligence, consciousness, and thought. This awe-inspiring organ interprets sensory information, formulates responses, and orchestrates a symphony of commands that govern our actions and behaviors.

The spinal cord, a long, slender structure protected by the spinal column, is a vital conduit between the brain and the rest of the body. It relays signals to and from the brain, enabling coordinated movements, reflexes, and sensory experiences.

The peripheral nervous system extends its delicate tendrils outward from the CNS, reaching every corner of the body. This vast network includes sensory neurons that transmit information from sensory organs to the CNS and motor neurons that carry instructions from the CNS to muscles and glands.

The nervous system's fundamental unit is the neuron, a specialized cell that transmits electrical signals. Neurons communicate through synapses, microscopic junctions where signals are relayed from one neuron to the next.

Through this network, the nervous system ensures swift and precise communication, allowing us to perceive the world, react to stimuli, and maintain homeostasis. It enables us to experience sensations like touch, taste, smell, sight, and sound, shaping our perceptions and enriching our understanding of the environment.

But the nervous system's role goes far beyond sensory perception. It regulates and coordinates vital bodily functions, including heartbeat, respiration, digestion, and secretion. It governs the movement of our muscles, allowing us to walk, run, dance, and perform intricate tasks with astonishing agility.

The nervous system is also responsible for higher cognitive functions like memory, learning, reasoning, and emotions. It shapes our personalities, influences our behaviors, and enables the formation of deep connections with others.

As we venture into this exploration, we'll plunge into the fascinating aspects of the nervous system. Our journey will introduce us to its complex structures, such as neurons, glial cells, and many neurotransmitters facilitating intercellular communication. Delving further, we will decode the subtle mechanisms that govern sensory perception, motor control, and the delicate balance maintained by the autonomic nervous system's sympathetic and parasympathetic divisions.

Chapter 2:
Circulatory System

Introduction to the Circulatory System

Welcome to the awe-inspiring realm of the circulatory system, a remarkable network of vessels and organs that tirelessly transports life-giving substances throughout the human body. From the heart's rhythmic beating to the pathways of blood vessels, this chapter will unravel the mysteries of one of the most vital systems in our spectacular human bodies.

The circulatory system, also known as the cardiovascular system, is pivotal in maintaining the body's equilibrium and sustaining life. It is responsible for the circulation of blood, which acts as a transportation system, carrying essential nutrients, oxygen, hormones, and immune cells to every nook and cranny of our bodies.

At the heart of this remarkable system is the muscular organ that shares its name, the heart. The heart, nestled within the thoracic cavity, is a powerful pump that propels blood throughout the body with each rhythmic contraction. It is the driving force behind the continuous flow of blood, ensuring that vital substances reach every cell, tissue, and organ that depends on them.

The circulatory system has a complex blood vessel network that spans the entire body. Arteries, veins, and capillaries form an intricate web, carrying blood to and from various organs and tissues and facilitating the exchange of oxygen, nutrients, and waste products.

Embark with us on a captivating expedition through the boundless universe of the circulatory system. Witness how this sophisticated web flawlessly guarantees the delivery of life-essential elements to every nook and cranny of our bodies, offering the nourishment and oxygenation critical for peak functioning. Steel yourself to be astounded by this unique system's grand complexity and elegance, rhythmically throbbing to the beat of life. Together, let's decode the enigmas of the circulatory system, plunging into its profound depths and marveling at its remarkable capacity to sustain energy and foster vitality.

Components of the Circulatory System (Heart, Blood Vessels)

Within the multifaceted tapestry of the circulatory system, two key components reign supreme: the heart and the blood vessels. These remarkable structures work harmoniously to ensure the seamless flow of blood and the efficient distribution of life-sustaining substances throughout our bodies.

The Heart

At the epicenter of the circulatory system lies the heart, a marvel of muscular precision. This vital organ, roughly the size of a clenched fist, beats tirelessly, propelling blood through the vast network of blood vessels. It is truly the engine that keeps the circulatory system running smoothly.

The heart is divided into four chambers: two atria and two ventricles. The atria act as receiving chambers, while the ventricles function as the pumping powerhouses. These chambers work in perfect synchrony, allowing for the controlled and efficient circulation of blood.

The heart is equipped with valves to ensure unidirectional blood flow. These valves open and close with each heartbeat, allowing blood to move forward while preventing backflow. The coordinated contraction and relaxation of the heart's chambers, combined with the precise opening and closing of its valves, create the rhythmic symphony that drives circulation.

Blood Vessels

The circulatory system boasts a meticulous network of blood vessels that act as the conduits through which blood flows. These vessels come in three main types: arteries, veins, and capillaries, each playing a distinct role in the complex circulation web.

Arteries are the sturdy highways that carry oxygenated blood away from the heart and into the body's tissues. Their thick and elastic walls withstand the forceful contractions of the heart, ensuring that blood reaches even the farthest reaches of the body. Arteries gradually branch out into smaller vessels called arterioles, which further divide into the smallest vessels known as capillaries.

With their wafer-thin walls, capillaries form an interconnected network that connects arteries to veins. It is within these microscopically narrow passages that the exchange of oxygen, nutrients, and waste products occurs. Capillaries allow vital substances to diffuse out of the bloodstream and into the surrounding tissues, nourishing and facilitating their functions. Simultaneously, waste products, such as carbon dioxide, return to the capillaries to be carried away.

Veins, the return journey of the circulatory system, carry deoxygenated blood back to the heart. They gradually merge into larger vessels and eventually converge into two major veins, the superior and inferior vena cava, which return blood to the heart's right atrium. Veins have thinner walls than arteries and rely on valves to prevent the backward flow of blood, aiding its upward journey against gravity.

These components—heart, arteries, capillaries, and veins— collaborate in perfect synergy, creating a finely tuned system that ensures the delivery of oxygen, nutrients, hormones, and immune cells to every part of our bodies. Together, they navigate the vast landscape of the circulatory system, forging pathways of life and vitality.

Pulmonary and Systemic Circuits

Within the vast network of the circulatory system, two essential circuits work in tandem to ensure the proper functioning of our bodies: the pulmonary circuit and the systemic circuit. These circuits act as interconnected pathways, guiding blood flow to distinct destinations and serving different purposes.

Pulmonary Circuit

The pulmonary circuit, starting from the right side of the heart, involves transporting deoxygenated blood from various body parts. The blood enters the superior and inferior vena cava and is transferred to the lungs for oxygenation. Join us as we trace the path of this extraordinary circuit, observing the dramatic metamorphosis of blood as it undergoes its vital stages of transformation. From the right atrium, the blood flows into the right ventricle, which then contracts, propelling the blood through the pulmonary artery.

The pulmonary artery, the only artery in the body that carries deoxygenated blood, divides into smaller branches, subdivided into tiny capillaries that envelop the walls of the lung's air sacs, known as alveoli. Here, a remarkable exchange takes place. Carbon dioxide, a waste product of cellular respiration, diffuses out of the capillaries and into the alveoli. At the same time, oxygen from the inhaled air enters the capillaries and binds to hemoglobin within red blood cells.

Oxygenated blood then travels back to the heart, but this time, it enters the left side of the heart through the pulmonary veins. The left atrium receives the oxygenated blood and passes it into the left ventricle. With a powerful contraction, the left ventricle propels the oxygen-rich blood into the aorta, the largest artery in the body.

Systemic Circuit

The systemic circuit delivers oxygenated blood to all the organs, tissues, and cells throughout the body, nourishing them and providing them with the essential substances they need to thrive. Let us venture into this vast and detailed circuit, exploring the pathways it takes and the vital destinations it reaches.

The aorta emerges from the left ventricle, acting as the systemic circuit's main thoroughfare. The aorta branches out, forming smaller arteries that supply oxygenated blood to various body regions. These arteries further divide into arterioles, eventually leading to a compounded network of capillaries that permeate nearly every tissue and organ.

The exchange of oxygen, nutrients, and waste products occurs within the capillaries. Oxygen and nutrients diffuse out of the capillaries and into the surrounding tissues, providing nourishment and energy for cellular processes.

As blood continues its journey through the capillaries, it gradually converges into venules, which merge to form veins. These veins collect the deoxygenated blood and carry it back to the heart, specifically to the right atrium, initiating the cycle again.

Through the coordinated efforts of the pulmonary and systemic circuits, the circulatory system ensures the continuous flow of oxygenated blood, vital nutrients, and essential substances to every nook and cranny of our bodies. The pulmonary circuit facilitates the exchange of gases, replenishing oxygen supplies and eliminating carbon dioxide. In contrast, the systemic course nourishes our cells, enabling growth, repair, and optimal functioning.

The Function of Blood and Its Role in Transportation

Blood, the life-sustaining fluid coursing through our bodies, plays a vital role in transporting essential substances, ensuring the proper functioning of our organs and tissues. A meticulous symphony of cells and molecules collaborates within its crimson currents to deliver oxygen, nutrients, hormones, and waste products to their intended destinations.

Oxygen Transport

Transporting oxygen from the lungs to every cell in the body is one of the fundamental roles carried out by blood. Oxygen binds to hemoglobin, a protein found in red blood cells, forming a dynamic partnership for efficient oxygen delivery. As blood flows through the lungs during inhalation, oxygen molecules diffuse across the delicate alveolar-capillary membrane, binding to hemoglobin within red blood cells. As the circulatory system transports them, red blood cells carrying a load of oxygen embark on a remarkable journey to tissues and organs hungry for this vital element. Through this mechanism, blood ensures that cells receive the oxygen necessary for cellular respiration, which generates energy to fuel our bodily functions.

Nutrient Transport

Blood serves as a crucial conduit for delivering vital nutrients to our cells. From the nutrients derived from the foods we consume to the molecules synthesized by our organs, blood carries an array of sustenance to the hungry cells that depend on them. Through the systemic circuit, arteries, and capillaries transport nutrients such as glucose, amino acids, fatty acids, vitamins, and minerals to various tissues and organs. These nutrients are absorbed from the digestive system, extracted by the liver, and then released into the bloodstream to nourish cells. This intricate transport system ensures that our cells receive the building blocks and energy sources they need to sustain life.

Hormone Transport

Hormones, the body's chemical messengers, rely on blood to reach their target organs and elicit specific responses. Endocrine glands, such as the pituitary, thyroid, adrenal glands, and others, release hormones into the bloodstream, where they hitch a ride to their intended destinations. Blood acts as the courier, transporting hormones to distant organs and tissues, binding to specific receptors, and initiating essential physiological processes. Whether regulating metabolism, growth, reproduction, or maintaining homeostasis, blood facilitates the timely delivery of hormones, ensuring the smooth coordination of countless bodily functions.

Waste Product Removal

Just as blood transports vital substances, it also plays a crucial role in removing waste products from our cells and tissues. Metabolic byproducts, such as carbon dioxide and other waste molecules, accumulate as cells carry out their activities. Blood serves as a scavenger, collecting these waste products and carrying them away to organs responsible for their elimination. Carbon dioxide, a waste product of cellular respiration, is transported by the blood from the tissues to the lungs, where it is exhaled. Similarly, other waste products are transported to the kidneys, liver, and other organs for processing and removal from the body. Through this process, blood maintains the delicate balance necessary for optimal health.

The multifaceted nature of blood and its role as a transportation system is nothing short of astonishing. It serves as a lifeline, nourishing our cells, removing waste, and facilitating essential communication within our bodies. The complicated interplay of red blood cells, plasma, and various other components ensures the seamless delivery of oxygen, nutrients, hormones, and waste to their destinations.

Common Diseases and Conditions Related to the Circulatory System

While the circulatory system is a remarkable network that sustains our bodies, it is not impervious to ailments. Various diseases and conditions can affect this vital system's delicate balance and functioning. Understanding these common disorders is crucial for recognizing their symptoms, seeking appropriate medical care, and adopting preventive measures. Let us explore some of the most prevalent diseases and conditions related to the circulatory system.

Hypertension (High Blood Pressure)

Hypertension is a condition characterized by persistently elevated blood pressure levels. It places excessive strain on the blood vessels, heart, and other organs, increasing the risk of serious complications such as heart disease, stroke, and kidney problems. Hypertension often has no symptoms but can be detected through regular blood pressure measurements. Lifestyle modifications, medication, and managing underlying health conditions are commonly employed to control hypertension and minimize its impact on the circulatory system.

Coronary Artery Disease (CAD)

Coronary artery disease occurs when plaque, consisting of cholesterol and other substances, builds up in the arteries supplying the heart with blood. Over time, this can narrow or block the arteries, restricting blood flow to the heart muscle. CAD can lead to angina (chest pain), heart attacks, and heart failure. Lifestyle changes, medications, and medical procedures such as angioplasty or bypass surgery are often employed to manage CAD and restore blood flow to the heart.

Stroke

A stroke occurs when the blood supply to a part of the brain is interrupted or reduced, leading to brain cell damage and potentially permanent neurological deficits. Ischemic stroke, caused by a blockage in a blood vessel supplying the brain, and hemorrhagic stroke, caused by bleeding in the brain, are the two main types. Prompt medical attention is crucial in managing strokes to minimize damage and maximize recovery. Rehabilitation and preventive measures are essential in reducing the risk of recurrent strokes.

Peripheral Artery Disease (PAD)

Peripheral artery disease affects the arteries outside the heart and brain, typically those supplying the limbs. It is characterized by the narrowing or blockage of these arteries, leading to reduced blood flow to the legs and arms. Symptoms may include leg pain, cramping, and slow wound healing. Lifestyle changes, medication, and medical procedures are employed to manage PAD and improve blood circulation to the affected limbs.

Deep Vein Thrombosis (DVT)

Deep vein thrombosis occurs when a blood clot forms in one of the deep veins, commonly in the legs. DVT can be caused by prolonged immobility, surgery, or certain medical conditions. Failure to receive treatment may result in the dislodgment of the clot, which can then migrate to the lungs, leading to a critical condition known as pulmonary embolism, which poses a potential threat to life. Medication, compression stockings, and lifestyle modifications are employed to prevent and treat DVT.

Atherosclerosis

Atherosclerosis is a condition characterized by the buildup of plaque inside the arteries. This plaque consists of cholesterol, fatty substances, calcium, and other materials that narrow and harden the arteries, restricting blood flow to vital organs. Atherosclerosis can affect various arteries in the body and is a major contributor to heart disease, stroke, and peripheral artery disease. Lifestyle changes, medication, and medical procedures may be employed to manage atherosclerosis and prevent complications.

These are just a few diseases and conditions that can affect the circulatory system. Maintaining a healthy lifestyle, managing risk factors such as high blood pressure and cholesterol levels, and seeking timely medical care are essential for preserving the health of this system. Understanding these common disorders can proactively protect our circulatory system and ensure optimal functioning, allowing us to lead vibrant and fulfilling lives.

Chapter 3:
Digestive System

The digestive system is a complex network of organs and processes that enables our bodies to break down food, absorb essential nutrients, and eliminate waste products. It is vital in providing the fuel and building blocks necessary for our growth, development, and overall well-being. Let's embark on the fascinating world of the digestive system and explore its key components and functions.

The digestive system begins in the mouth, where food is initially ingested and broken down into smaller particles through chewing and mixing with saliva. The food then travels down the esophagus, a muscular tube that connects the mouth to the stomach, through peristalsis—a series of coordinated contractions that propel the food forward.

Upon reaching the stomach, the food encounters gastric juices and enzymes, which help break it down further and facilitate the digestion of proteins. The stomach's muscular walls churn and mix the food, forming a partially digested semi-liquid substance called chyme. This chyme is gradually released into the small intestine, where most nutrient absorption occurs.

The small intestine, a long and intricately folded tube, consists of three segments: the duodenum, jejunum, and ileum. Here, the breakdown of carbohydrates, proteins, and fats reaches its completion with the help of various digestive enzymes produced by the pancreas and intestinal cells. Nutrients, such as amino acids, glucose, and fatty acids, are then absorbed into the bloodstream and transported to the body's cells for energy production, growth, and repair.

The remaining undigested material, including fiber, passes into the large intestine, also known as the colon. In the colon, water is reabsorbed, and the waste products are formed into solid feces. The feces are stored in the rectum until eliminated through the anus during a bowel movement.

Throughout this complex process, the digestive system relies on the coordinated actions of various organs, including the liver, gallbladder, and pancreas, which play crucial roles in producing and delivering digestive enzymes, bile, and other substances necessary for proper digestion.

Understanding the functioning of the digestive system helps us appreciate the incredible complexity and efficiency with which our bodies process the food we consume. By optimizing our dietary choices, maintaining a balanced lifestyle, and addressing any digestive concerns, we can support the healthy functioning of our digestive system and promote overall wellness.

Organs Involved in Digestion

The digestive system relies on a series of organs working harmoniously to break down food and extract nutrients. Let's explore the key organs involved in the process of digestion.

Mouth

The journey of digestion begins in the mouth, where food is ingested, and initial mechanical breakdown occurs through chewing. The saliva secreted by salivary glands helps moisten the food, making it easier to swallow and initiating the chemical analysis of starches with the enzyme amylase.

Esophagus

Once the food is swallowed, it travels down the esophagus, a muscular tube connecting the mouth to the stomach. Rhythmic contractions propel the food downward through peristalsis, ensuring its safe passage to the stomach.

Stomach

The stomach is a muscular organ that acts as a reservoir for food and aids in further mechanical and chemical digestion. It secretes gastric juices containing hydrochloric acid and enzymes, such as pepsin, which break down proteins. The stomach's muscular contractions mix the food with gastric juices, forming chyme. This partially digested mixture is gradually released into the small intestine.

Small Intestine

The small intestine is a long, coiled tube where most nutrient absorption occurs. It consists of three segments: the duodenum, jejunum, and ileum. The small intestine receives digestive enzymes from the pancreas and bile from the liver and gallbladder to aid in the breakdown of carbohydrates, proteins, and fats. The small intestine's inner lining is covered in finger-like projections called villi, which greatly increase the surface area for nutrient absorption into the bloodstream.

Large Intestine

As the remaining undigested and unabsorbed food reaches the large intestine, water and electrolytes are reabsorbed, forming feces. The large intestine also houses beneficial bacteria that aid in the breakdown of undigested fibers and produce certain vitamins.

These organs work in a coordinated manner to break down food into smaller molecules, ensuring optimal absorption and utilization of nutrients by the body. Understanding the roles of these organs in the digestive system provides us with a deeper appreciation for the intricacies involved in turning food into usable energy and building blocks for our bodies.

Role of Accessory Organs (Liver, Pancreas, Gallbladder)

In addition to the primary organs involved in digestion, the digestive system relies on several accessory organs that play crucial roles in the process. These organs work together to ensure the efficient breakdown and absorption of nutrients. Let's dive into the remarkable functions of the liver, pancreas, and gallbladder.

Liver

The liver is the largest internal organ in the human body and performs multiple functions vital to digestion. One of its primary roles is the production of bile, a greenish-yellow fluid that aids in the digestion and absorption of fats. Bile is stored and concentrated in the gallbladder before being released into the small intestine. The liver also metabolizes nutrients, detoxifies harmful substances, stores vitamins, and minerals, and produces important proteins necessary for blood clotting and immune function.

Pancreas

The pancreas is an endocrine and exocrine gland that releases hormones directly into the bloodstream and digestive enzymes into the small intestine. The pancreatic enzymes, including amylase, lipase, and proteases, are crucial for breaking down carbohydrates, fats, and proteins. These enzymes are released into the small intestine to further aid food digestion. Additionally, the pancreas produces insulin and glucagon, hormones regulating blood sugar levels.

Gallbladder

The gallbladder is a small, pear-shaped organ located beneath the liver. Its primary function is to store and concentrate bile produced by the liver. When fatty foods enter the small intestine, the gallbladder contracts, releasing bile into the digestive tract. Bile helps emulsify fats, breaking them into smaller droplets and facilitating their digestion by enzymes.

The coordinated actions of the liver, pancreas, and gallbladder are crucial for optimal digestion and nutrient absorption. These accessory organs work in harmony with the primary organs to ensure the effective breakdown of food and the absorption of essential nutrients into the bloodstream.

Common Digestive Disorders and Their Impact on Health

The digestive system is a complex network of organs and processes that combine to break down food and extract nutrients. However, this system can sometimes be susceptible to various disorders that significantly impact an individual's health and well-being. Let's explore some common digestive disorders and their effects on the body.

Gastroesophageal Reflux Disease (GERD)

GERD is characterized by the backward flow of stomach acid into the esophagus. This can cause symptoms such as heartburn, chest pain, regurgitation, and difficulty swallowing. Over time, untreated GERD can lead to complications like esophageal inflammation, ulcers, and an increased risk of esophageal cancer.

Peptic Ulcer Disease

Peptic ulcers are open sores that develop on the stomach lining, small intestine, or esophagus. These ulcers are often caused by a bacterial infection called Helicobacter pylori or prolonged nonsteroidal anti-inflammatory drugs (NSAIDs) use. Symptoms may include abdominal pain, bloating, nausea, and vomiting. Peptic ulcers can lead to serious complications such as bleeding, perforation, or obstruction without proper treatment.

Inflammatory Bowel Disease (IBD)

IBD is an umbrella term for chronic conditions that cause inflammation in the digestive tract. The two primary forms of IBD are Crohn's disease and ulcerative colitis. Both conditions can lead to symptoms such as abdominal pain, diarrhea, rectal bleeding, fatigue, and weight loss. IBD is a lifelong condition that requires ongoing management and can significantly impact a person's quality of life.

Irritable Bowel Syndrome (IBS)

IBS is a functional digestive system disorder characterized by symptoms like abdominal pain, bloating, constipation, and diarrhea. While the exact cause of IBS is unknown, triggers may include certain foods, stress, and hormonal changes. Although IBS doesn't cause permanent damage to the digestive tract, it can greatly affect a person's daily life and well-being.

Celiac Disease

Celiac disease is an autoimmune condition activated upon ingesting gluten, a protein prevalent in wheat, barley, and rye. When individuals with celiac disease consume gluten, their immune system responds by damaging the lining of the small intestine, leading to various symptoms such as abdominal pain, diarrhea, fatigue, and nutrient deficiencies. Adherence to a rigorous gluten-free diet is the sole treatment for celiac disease.

These are just a few examples of the many digestive disorders that individuals may encounter. Understanding these conditions and their impact on health is crucial for early detection, proper management, and overall well-being.

Chapter 4:
Respiratory System

The respiratory system is a remarkable network of organs and tissues that enables us to breathe, providing our bodies with the essential oxygen needed for survival. From the moment we take our first breath to the countless inhalations and exhalations throughout our lives, the respiratory system plays a vital role in sustaining our existence. In this chapter, we will embark on a fascinating journey into the workings of the respiratory system.

Importance of the Respiratory System

The respiratory system exchanges oxygen and carbon dioxide between our bodies and the surrounding environment. It ensures that oxygen is taken in and transported to every cell in the body while simultaneously removing carbon dioxide, a waste product of cellular metabolism. Without this continuous gas exchange process, our bodies would be deprived of the oxygen necessary for energy production. They would be overwhelmed by the buildup of toxic carbon dioxide.

Structure and Components

The respiratory system has several key components, each with a specific role in facilitating respiration. It includes the nose, nasal cavity, pharynx, larynx, trachea, bronchi, bronchioles, and lungs.

These structures work together seamlessly to carry out the processes of ventilation, gas exchange, and oxygen utilization within our bodies.

Mechanism of Breathing

Breathing, or pulmonary ventilation, is the process by which air is drawn into the lungs and expelled from the body. It involves the coordinated actions of the diaphragm, intercostal muscles, and various respiratory muscles. Understanding the mechanics of breathing and the factors that influence respiratory rate and depth is essential for comprehending the functioning of the respiratory system.

Gas Exchange and Transport

The exchange of gases, primarily oxygen, and carbon dioxide, occurs within the tiny air sacs called alveoli, which are found in the lungs. Here, oxygen from the inhaled air diffuses into the bloodstream. At the same time, carbon dioxide, produced as a waste product by our cells, moves from the blood into the alveoli to be exhaled. The respiratory system works with the circulatory system to efficiently transport gases to and from tissues throughout the body.

Respiratory Regulation

The respiratory system is tightly regulated to maintain homeostasis and ensure a constant supply of oxygen and carbon dioxide removal. The control of breathing is influenced by various factors, including neural mechanisms, chemical sensors, and feedback loops. These regulatory processes help adjust our breathing rate and depth in response to changing body oxygen and carbon dioxide levels.

As we immerse ourselves in the details of the respiratory system, we are bound to gain a deep admiration for its indispensable role in the continuation of human life. The forthcoming chapters will bring us a closer look at each part's structure and function, decode the processes of gas exchange and transport, and scrutinize prevalent respiratory disorders and their influence on our health. United in our pursuit, we will reveal the astounding intricacies of the respiratory system, thereby deepening our comprehension of its fundamental importance in our day-to-day existence.

Structures Involved in Respiration

Within the intricate web of the respiratory system, several key structures work harmoniously to facilitate the process of respiration.

The Nose

The nose serves as the primary entrance for air into the respiratory system. Its external portion, the nostrils, allows the intake of oxygen-rich air. As air enters the nasal cavity, it is filtered, humidified, and warmed by the mucous membranes and tiny hair-like structures called cilia. These mechanisms help protect the delicate respiratory system from foreign particles and ensure that the air we breathe is clean and suitable for the subsequent respiration processes.

The Trachea

The trachea, also known as the windpipe, is a tube-like structure that connects the larynx to the bronchi. It consists of a series of cartilaginous rings that provide structural support, preventing airway collapse during respiration. Within the trachea, specialized cells called goblet cells produce mucus, which helps to trap foreign particles and facilitate the removal of debris from the respiratory system.

The Lungs

The lungs are the central organs of the respiratory system, responsible for the exchange of gases between the air and the bloodstream. Protected by the rib cage, the lungs are two spongy, cone-shaped structures in the thoracic cavity. They are composed of millions of tiny air sacs called alveoli, where oxygen and carbon dioxide exchange occurs. The lungs are surrounded by a double-layered membrane called the pleura, which helps to reduce friction during breathing.

Bronchi and Bronchioles

Within the lungs, the trachea branches into two main bronchi, one leading to each lung. These bronchi further divide into smaller bronchioles, which extend deep into the lung tissue. Bronchioles are lined with smooth muscles that regulate airflow and play a crucial role in maintaining proper ventilation.

Alveoli

The alveoli are the microscopic air sacs where the exchange of gases occurs. These thin-walled structures provide an extensive surface area for the diffusion of oxygen from inhaled air into the bloodstream and the release of carbon dioxide into the exhaled air. The delicate alveolar walls are richly supplied with tiny blood vessels called capillaries, ensuring efficient gas exchange.

Mechanics of Breathing and Gas Exchange

The respiratory system's mechanics of breathing and gas exchange are fascinating processes that allow our bodies to constantly supply oxygen to our cells and remove carbon dioxide, maintaining a delicate balance vital for survival. Let's investigate the mechanisms involved in this remarkable journey.

When we inhale, the process begins with the contraction of the diaphragm, a dome-shaped muscle located at the base of the chest. Simultaneously, the intercostal muscles between the ribs contract, causing the ribcage to expand. These actions create negative pressure within the chest cavity, drawing air into the lungs through the airways.

Common Respiratory Conditions and Their Effects on Breathing

With its delicate balance of structures and processes, the respiratory system can be susceptible to various conditions that impact our breathing. Let's explore some of the most common respiratory disorders and their effects on the respiratory system.

Asthma

Asthma is a chronic condition characterized by inflammation and narrowing of the airways. This leads to recurring wheezing, shortness of breath, chest tightness, and coughing. During an asthma attack, the muscles surrounding the airways contract, causing further narrowing and making breathing difficult. Asthma can be triggered by various factors such as allergens, exercise, or respiratory infections.

Chronic Obstructive Pulmonary Disease (COPD)

COPD is a progressive lung disease that encompasses chronic bronchitis and emphysema. Chronic bronchitis involves inflammation and excessive mucus production in the airways, leading to coughing and difficulty breathing. Emphysema is characterized by damage to the alveoli, reducing the lungs' ability to efficiently exchange oxygen and carbon dioxide. COPD is often caused by long-term exposure to cigarette smoke or other irritants.

Pneumonia

Pneumonia is characterized by the inflammation of air sacs in either one or both lungs due to an infection. Bacteria, viruses, or fungi are potential culprits that can induce pneumonia. Pneumonia can lead to coughing, fever, chest pain, and difficulty breathing. In severe cases, it can significantly impair lung function and require medical intervention.

Allergic Rhinitis

Allergic rhinitis, known as hay fever, is an allergic reaction triggered by airborne substances such as pollen, dust mites, or pet dander. It causes inflammation of the nasal passages, leading to symptoms like sneezing, itching, nasal congestion, and a runny nose. These symptoms can affect breathing and overall respiratory comfort.

Sleep Apnea

Sleep apnea is a sleep disorder characterized by pauses in breathing or shallow breaths during sleep. It occurs when the muscles in the throat fail to keep the airway open, leading to brief interruptions in breathing. Sleep apnea can result in loud snoring, fragmented sleep, excessive daytime sleepiness, and reduced oxygen levels during sleep.

These are just a few respiratory conditions that can impact breathing and overall respiratory health. Understanding these conditions and their effects is essential for early detection, proper management, and respiratory well-being.

Chapter 5:
Nervous System

The nervous system is an intricate network of cells and tissues that plays a fundamental role in communication and coordination throughout the body. It is responsible for processing sensory information, initiating and regulating body movements, and maintaining homeostasis.

The nervous system can be divided into two primary components: the central nervous system (CNS) and the peripheral nervous system (PNS). The CNS consists of the brain and the spinal cord, which serve as the command center for the body's activities. The PNS includes the nerves that extend from the CNS to various body parts, enabling communication between the CNS and the rest of the body.

Within the nervous system, neurons and glial cells transmit and process information. Neurons are specialized cells that transmit electrical impulses, allowing for the rapid communication necessary for coordinated body functions. Glial cells provide support and protection to neurons, maintaining their environment and assisting in their functioning.

The nervous system can be further categorized into different divisions based on its functions. The somatic nervous system controls voluntary movements and relays sensory information, such as touch and temperature, to the CNS. In contrast, the autonomic nervous system regulates involuntary processes like heart rate, digestion, and breathing. The autonomic nervous system is further divided into the sympathetic and parasympathetic divisions, which work harmoniously to maintain balance in response to internal and external stimuli.

In addition to its vital role in controlling bodily functions, the nervous system also encompasses the special senses of vision, hearing, taste, and smell. The eyes, ears, tongue, and nose serve as sensory receptors, gathering information about the external environment and relaying it to the brain for interpretation.

Structure and Function of Neurons and Glial Cells

Neurons and glial cells are the essential building blocks of the nervous system, each playing a unique role in transmitting and processing information. Let's explore the structure and function of these remarkable cells.

Neurons, which are referred to as nerve cells, form the bedrock of the nervous system. They are specialized to transmit electrical impulses, or nerve impulses, to communicate and coordinate various functions within the body. Structurally, neurons have three main components: the cell body, dendrites, and axons.

The cell body, called the soma, contains the nucleus and other essential organelles. It serves as the control center of the neuron, responsible for maintaining cell functions and synthesizing important proteins necessary for its operation.

Dendrites are branching extensions that receive incoming signals from other neurons or sensory receptors. These signals, in the form of electrical impulses, travel across the dendrites and converge at the cell body. Dendrites are crucial in integrating and relaying information to the cell body.

Axons are long, slender projections that carry nerve impulses from the cell body to other neurons or target cells, such as muscles or glands. Covered by a protective myelin sheath, axons facilitate the rapid transmission of electrical signals over long distances. The myelin sheath is formed by specialized glial cells known as oligodendrocytes in the CNS and Schwann cells in the PNS.

At the end of axons are specialized structures called axon terminals or synaptic terminals. These terminals form connections, called synapses, with other neurons or target cells. Within the synapse, electrical impulses are converted into chemical signals that can then be transmitted to the next neuron or target cell.

Glial cells, or neuroglia, provide crucial support and protection to neurons. Although often overlooked, these cells are essential for properly functioning the nervous system. Glial cells outnumber neurons and come in several types, including astrocytes, oligodendrocytes, microglia, and ependymal cells.

Astrocytes are star-shaped cells that help maintain the chemical environment surrounding neurons. They regulate the concentration of ions and neurotransmitters, provide nutrients to neurons, and play a role in repairing damaged brain tissue.

Oligodendrocytes and Schwann cells produce the myelin sheath, insulating axons and enhancing the speed of nerve impulse transmission. In addition to their role in myelination, oligodendrocytes provide structural support to neurons in the CNS.

Microglia function as the immune cells of the nervous system. They survey the brain for pathogens or abnormal cells. They are activated in response to injury or infection, aiding tissue repair and regulating inflammation.

Ependymal cells line the fluid-filled spaces within the brain and spinal cord. They contribute to the production and circulation of cerebrospinal fluid, which provides cushioning and support to the nervous system.

Neurons and glial cells work together in a coordinated manner, allowing for the complex functions of the nervous system. Neurons transmit electrical impulses, while glial cells provide support and maintenance. This multifaceted partnership ensures the proper functioning of the nervous system and enables the transmission of information throughout the body.

Central Nervous System: The Brain and Spinal Cord

The central nervous system (CNS) comprises two vital components: the brain and the spinal cord.

The brain, often considered the body's command center, is a complex organ responsible for a wide range of functions, including cognition, emotions, memory, and voluntary movements. Structurally, the brain consists of distinct regions with unique roles and responsibilities.

The cerebral cortex, the brain's outer layer, plays a critical role in higher cognitive functions, such as perception, language, reasoning, and decision-making. It is divided into two hemispheres—left and right—interconnected by a bundle of nerve fibers called the corpus callosum.

Beneath the cerebral cortex lie deep structures known as the basal ganglia, which are involved in coordinating movement and regulating voluntary motor control. These structures work with the cerebral cortex and other regions to facilitate smooth and purposeful actions.

The limbic system, situated deep within the brain, is involved in emotions, memory formation, and motivation. It includes structures such as the amygdala, hippocampus, and hypothalamus, which regulate emotional responses, form memories, and control basic physiological functions like hunger, thirst, and sleep.

The brainstem, connecting the brain to the spinal cord, controls vital functions necessary for survival, such as breathing, heart rate, and blood pressure. It also serves as a relay center, transmitting sensory and motor information between the brain and the rest of the body.

Moving on to the spinal cord is a long, cylindrical bundle of nerve fibers extending from the base of the brain down through the spinal column. The spinal cord is a vital conduit for communication between the brain and the peripheral nervous system (PNS).

Within the spinal cord, sensory neurons transmit information from the body's periphery to the brain, allowing us to perceive sensations such as touch, pain, and temperature. Motor neurons, on the other hand, carry signals from the brain to the muscles and glands, enabling voluntary movements and regulating physiological processes.

The brain and spinal cord form a complex and interconnected network, facilitating communication and coordination throughout the body. These organs work with the peripheral nervous system to process sensory information, initiate appropriate responses, and maintain homeostasis.

In the subsequent sections, we will explore the subtle functions of different brain regions, the remarkable capabilities of the spinal cord, and how they collectively contribute to our perception, behavior, and overall well-being. The central nervous system represents the pinnacle of evolutionary development, allowing us to navigate the world and experience the wonders of human consciousness.

Peripheral Nervous System and Its Subdivisions

Beyond the central nervous system, the human body houses an interconnected network known as the peripheral nervous system (PNS). The PNS extends throughout the body, connecting the central nervous system to various organs, muscles, and sensory receptors. Let's review the subdivisions of the peripheral nervous system and its essential functions.

Somatic Nervous System (SNS)

The somatic nervous system is responsible for voluntary movements and sensory perception. It enables us to interact with the external environment and consciously control our actions. Motor neurons within the SNS transmit signals from the central nervous system to skeletal muscles, initiating purposeful movements. On the other hand, sensory neurons carry information from sensory organs and receptors to the central nervous system, providing us with sensations like touch, temperature, and pain.

Autonomic Nervous System (ANS)

The autonomic nervous system operates involuntarily, regulating vital functions necessary for sustaining life. It controls the activities of internal organs, glands, and smooth muscles. The ANS consists of two main subdivisions: the sympathetic nervous system and the parasympathetic nervous system.

Sympathetic Nervous System

The sympathetic nervous system is responsible for the body's "fight-or-flight" response. It prepares the body for action in response to perceived threats or stressors. Activating the sympathetic nervous system leads to increased heart rate, elevated blood pressure, dilated pupils, and the release of stress hormones like adrenaline, mobilizing the body's resources for immediate action.

Parasympathetic Nervous System

In contrast to the sympathetic system, the parasympathetic nervous system promotes relaxation and restoration. It counterbalances the effects of the sympathetic system, facilitating rest, digestion, and recovery. Activation of the parasympathetic system slows the heart rate, enhances digestion, and conserves energy.

The coordination between the sympathetic and parasympathetic systems ensures a delicate balance in maintaining homeostasis, adapting to different situations, and responding appropriately to internal and external stimuli.

Enteric Nervous System (ENS)

The enteric nervous system is a unique division of the peripheral nervous system that resides within the walls of the gastrointestinal tract. It functions independently and controls digestion, nutrient absorption, and food movement through the intestines. The ENS contains an extensive network of neurons and is crucial in regulating gastrointestinal functions, such as peristalsis, secretion, and blood flow within the digestive system.

The peripheral nervous system, including the somatic, autonomic, and enteric divisions, works in harmony with the central nervous system to ensure the proper functioning of the entire body. It enables us to interact with the world, maintain physiological balance, and adapt to changing environments.

Role of the Nervous System in Sensory Perception and Motor Control

The nervous system, encompassing the central and peripheral components, plays a crucial role in our ability to perceive and interact with the world. A complex network of neurons and processes enables sensory perception and motor control, allowing us to experience sensations and initiate purposeful movements. Let's uncover the remarkable functions of the nervous system in sensory perception and motor control.

Sensory Perception

Sensory perception is how we receive and interpret information from our environment through various sensory modalities. The nervous system enables us to perceive and make sense of sensations such as touch, sight, hearing, taste, and smell. Here's an overview of how the nervous system contributes to sensory perception:

Sensory Receptors

Specialized sensory receptors throughout the body detect stimuli from the environment or within the body. These receptors convert different forms of energy, such as light, sound, or pressure, into electrical signals called nerve impulses.

Sensory Neurons

Nerve impulses generated by sensory receptors travel along sensory neurons, which transmit the signals to the central nervous system for processing. Each sensory modality has dedicated pathways that relay information to specific brain regions, where it is interpreted and integrated into meaningful perceptions.

Sensory Processing

In the brain, sensory information undergoes complex processing and interpretation. Different brain regions analyze and combine sensory inputs, allowing us to recognize objects, perceive depth, differentiate between colors, and interpret the qualities of various stimuli. This meticulous process ultimately forms our perception of the world.

Motor Control

Motor control refers to the ability to initiate and regulate voluntary movements. The nervous system is fundamental in coordinating and executing precise motor actions.

Motor Cortex

The motor cortex, located in the brain's frontal lobe, is responsible for planning, initiating, and controlling voluntary movements. It sends signals to the appropriate muscles through the spinal cord, enabling purposeful actions.

Motor Neurons

Motor neurons in the spinal cord and brainstem transmit signals from the motor cortex to the muscles. These signals, known as motor commands, trigger muscle contractions and coordinated movements.

Neuromuscular Junction

At the point where motor neurons connect with muscle fibers, known as the neuromuscular junction, neurotransmitters are released, initiating muscle contraction. This process allows for precise control of movements and the execution of complex actions.

Feedback Mechanisms

During movement, the nervous system continuously receives sensory feedback from muscles, joints, and other body parts. This feedback lets the brain monitor and adjust motor commands, ensuring coordinated and accurate movements.

The intricate interplay between sensory perception and motor control showcases the remarkable capabilities of the nervous system. By integrating sensory information and translating it into appropriate motor responses, the nervous system enables us to navigate our environment, interact with objects, and engage in activities essential to our daily lives.

Neurological Disorders and Their Impact on the Body

Neurological disorders are conditions that affect the structure or function of the nervous system, resulting in disruptions to normal brain, spinal cord, and peripheral nerve activity. These disorders can profoundly affect the body's functioning, leading to many symptoms and impairments. Let's explore some common neurological disorders and their impact on the body.

Alzheimer's Disease

Alzheimer's is a progressive neurodegenerative disorder characterized by memory loss, cognitive decline, and behavioral changes. It primarily affects the brain, causing a gradual deterioration of brain cells and the formation of abnormal protein structures. As the disease progresses, individuals may experience difficulties with memory, thinking, language, and daily activities.

Parkinson's Disease

Parkinson's disease is a chronic neurodegenerative disorder that affects movement. It occurs due to the loss of dopamine-producing cells in the brain, leading to motor symptoms such as tremors, rigidity, bradykinesia (slowness of movement), and postural instability. Non-motor symptoms, including depression, sleep disturbances, and cognitive impairment, can also manifest.

Multiple Sclerosis

Multiple sclerosis (MS) is characterized by an autoimmune response targeting the central nervous system. In MS, the immune system mistakenly attacks the protective covering of nerve fibers, disrupting communication between the brain and the rest of the body. This results in many symptoms, including fatigue, muscle weakness, coordination problems, sensory disturbances, and cognitive impairments.

Epilepsy

Epilepsy is a neurological disorder characterized by recurrent seizures. Seizures occur due to abnormal electrical activity in the brain, leading to temporary disruptions in normal brain function. The symptoms and severity of seizures can vary widely, ranging from momentary lapses of awareness to convulsive movements and loss of consciousness.

Stroke

A stroke occurs when the blood supply is disrupted to the brain, leading to brain cell damage. The effects of a stroke depend on the area of the brain affected and the extent of the damage. Common symptoms include sudden weakness or paralysis on one side of the body, speech difficulties, vision problems, and cognitive impairments.

Migraine

Migraine, a neurological disorder, manifests as recurring headaches and additional symptoms like nausea, sensitivity to light and sound, and visual disturbances. Migraines can significantly impact a person's quality of life. They may be triggered by various factors, including hormonal changes, certain foods, stress, or environmental factors.

Amyotrophic Lateral Sclerosis (ALS)

ALS, commonly called Lou Gehrig's disease, is a progressive neurodegenerative condition that impacts the nerve cells in the brain and spinal cord. It leads to the degeneration of motor neurons, resulting in muscle weakness, loss of coordination, and eventually paralysis. ALS affects voluntary muscle control, impacting speech, swallowing, and mobility.

These are just a few examples of the numerous neurological disorders, each with unique characteristics and impact on the body. Neurological disorders can affect various aspects of physical and cognitive functioning, leading to mobility, communication, cognition, and overall well-being challenges.

Understanding these disorders and their effects is crucial for developing effective treatments, improving patient care, and enhancing the quality of life for individuals with neurological conditions. Ongoing research and advancements in medical science offer hope for improved management and potential breakthroughs in neurological disorders.

Chapter 6:
Reproductive System

The reproductive system is a fascinating and essential aspect of the human body, responsible for creating new life and continuing the human species. It encompasses a complex network of organs, hormones, and processes that work together to facilitate reproduction. In this chapter, we're set to explore the fascinating aspects of the reproductive system, diving deep into its exceptional functionalities.

The reproductive system serves two primary purposes: the production of gametes (sperm and eggs) and the provision of a supportive environment for developing and nurturing offspring. It is divided into two distinct systems: the male reproductive system and the female reproductive system, each with its unique structures and functions.

Throughout this chapter, we will examine the reproductive anatomy, hormonal regulation, and the intricacies of the reproductive processes in both males and females. We will explore the development of reproductive organs, the production of gametes, and the journey of fertilization leading to the creation of new life.

Additionally, we will discuss the various factors that influence reproductive health, such as hormonal balance, fertility, and contraception. Understanding these factors is crucial for individuals and couples planning to start a family or seeking to maintain their reproductive well-being.

We will also address common reproductive disorders and conditions impacting fertility and reproductive function. By shedding light on these conditions, we aim to provide insights into their causes, symptoms, and available treatments, offering a comprehensive understanding of reproductive health issues.

The reproductive system is vital for the continuation of the human species. It is deeply intertwined with personal identity, emotions, and relationships. We will explore the psychological and social aspects of reproduction, including the impact of reproductive choices and the importance of reproductive education and healthcare.

Female Reproductive Anatomy and Physiology

The complex network of organs within the female reproductive system collaboratively enables the reproductive process and nurtures the growth of new life. Understanding the female reproductive anatomy and physiology is crucial for appreciating the remarkable capabilities of the female body.

A combination of internal and external organs can be found within the female reproductive system. Externally, the primary structures include the mons pubis, labia majora, labia minora, clitoris, and vaginal opening. The mons pubis is a fatty tissue pad covering the pubic bone. The labia majora and labia minora are folds of skin that protect the vaginal and urethral openings. The clitoris, located at the top of the labia minora, is a sensitive organ involved in sexual pleasure. The vaginal opening serves as the entrance to the reproductive tract.

Internally, the female reproductive system comprises several key organs. The main organ is the uterus, also known as the womb, responsible for nurturing and supporting the developing embryo and fetus during pregnancy. The uterus is connected to the ovaries by the fallopian tubes, which transport eggs from the ovaries to the uterus. The ovaries on each side of the uterus produce eggs (ova) and release hormones, including estrogen and progesterone, which regulate the menstrual cycle and support pregnancy.

The cervix is the lower, narrow portion of the uterus that connects it to the vagina. During childbirth, the cervix dilates to allow the baby's passage from the uterus into the birth canal. The vagina is a muscular tube that serves as the birth canal and is also the site for sexual intercourse. It connects the cervix to the external genitalia.

The female reproductive system undergoes cyclic changes known as the menstrual cycle. The menstrual cycle involves the release of an egg from one of the ovaries, known as ovulation, and the preparation of the uterus for the possible implantation of a fertilized egg. If fertilization does not occur, the lining of the uterus is shed during menstruation, marking the start of a new menstrual cycle.

Hormones play a vital role in regulating the female reproductive system. The hypothalamus in the brain releases gonadotropin-releasing hormone (GnRH), which stimulates the pituitary gland to produce follicle-stimulating hormone (FSH) and luteinizing hormone (LH). FSH stimulates the growth and maturation of follicles in the ovaries, each containing an egg. LH triggers ovulation and stimulates the production of estrogen and progesterone.

The female reproductive system is a phenomenal and sophisticated entity, capable of fostering the miracle of new life. Comprehending its structure and functioning is essential for preserving women's health, ensuring fertility, and thoughtful family planning. By understanding the subtleties within the female reproductive system, we can cultivate a profound respect for its awe-inspiring potential and the significant part it performs in the wonder of human reproduction.

Male Reproductive Anatomy and Physiology

The male reproductive system is a complex network of organs and structures that work together to produce, store, and deliver sperm for reproduction. Understanding the male reproductive anatomy and physiology is key to comprehending the processes involved in male fertility and reproduction.

The male reproductive system consists of both internal and external organs. Externally, the primary structures include the penis and the scrotum. The penis serves as the organ of copulation and delivers sperm into the female reproductive tract during sexual intercourse. The scrotum, a pouch of skin below the penis, houses the testicles and helps regulate the temperature required for optimal sperm production.

Internally, the male reproductive system comprises the testes, epididymis, vas deferens, seminal vesicles, prostate gland, and bulbourethral glands. The testes within the scrotum are responsible for producing sperm and testosterone, the primary male sex hormone. The epididymis, a coiled tube connected to each testicle, acts as a storage and maturation site for sperm.

The vas deferens is a muscular tube that carries mature sperm from the epididymis to the seminal vesicles, which mixes with seminal fluid to form semen. The seminal vesicles contribute fructose-rich fluid that provides energy for the sperm. The prostate gland produces a milky fluid that enhances sperm motility and neutralizes acidity in the female reproductive tract. The bulbourethral glands secrete a clear lubricating fluid that aids in the passage of semen during ejaculation.

The male reproductive system is regulated by a complex interplay of hormones. The hypothalamus in the brain releases gonadotropin-releasing hormone (GnRH), which stimulates the pituitary gland to produce follicle-stimulating hormone (FSH) and luteinizing hormone (LH). FSH promotes sperm production (spermatogenesis) in the testes, while LH stimulates testosterone production.

Spermatogenesis is a continuous process in which immature sperm cells undergo several stages of development to become mature, motile sperm. The mature sperm are then released into the epididymis for storage until ejaculation occurs.

During sexual arousal, the male reproductive system undergoes physiological changes. These changes include the engorgement of blood vessels in the penis, resulting in an erection, and the contraction of various muscles to propel semen through the urethra during ejaculation.

Understanding the male reproductive anatomy and physiology is essential for comprehending fertility issues, reproductive health, and preventing and treating male reproductive disorders. By gaining insights into the intricacies of the male reproductive system, we can develop a deeper appreciation for its remarkable capabilities and its crucial role in human reproduction.

Puberty, Menstruation, and Fertility

Puberty marks a significant milestone in both males and females, representing the onset of sexual maturation and the potential for reproduction. It is a transformative period characterized by physical, hormonal, and emotional changes. During puberty, the male and female reproductive systems undergo remarkable development, leading to distinct processes related to fertility.

Puberty typically begins between 9 and 14 in both sexes, although the exact timing can vary among individuals. It is triggered by the release of hormones from the hypothalamus in the brain, specifically gonadotropin-releasing hormone (GnRH). GnRH signals the pituitary gland to release follicle-stimulating hormone (FSH) and luteinizing hormone (LH), stimulating the testes in males and the ovaries in females to produce sex hormones.

One of the most noticeable changes during puberty is the development of secondary sexual characteristics. In males, these include the growth of facial and body hair, deepening of the voice, and increased muscle mass. In females, the changes involve the development of breasts, the widening of hips, the appearance of pubic and underarm hair, and the onset of menstruation.

Menstruation, or the menstrual cycle, is a natural process exclusive to females in which the lining of the uterus is shed through the vagina. It is a monthly occurrence that typically lasts around 28 days, although cycle lengths can vary. The menstrual cycle is regulated by the fluctuation of hormones, primarily estrogen and progesterone.

During each menstrual cycle, an egg is released from one of the ovaries in ovulation. If the released egg is fertilized by sperm, it may implant in the uterus, resulting in pregnancy. If fertilization does not occur, the lining of the uterus is shed, resulting in menstrual bleeding. This process does not appear in males.

Fertility, or the ability to conceive and bear children, typically begins after puberty in both males and females. It is important to note that fertility varies among individuals and can be influenced by various factors, including hormonal balance, overall health, and lifestyle choices. Understanding one's reproductive system and fertility patterns can empower family planning and reproductive health.

Puberty, menstruation, and fertility are significant aspects of the male and female reproductive journeys. They represent the biological mechanisms that enable individuals to conceive and nurture new life. By understanding and embracing these processes, individuals can make informed decisions about their reproductive health, contraception, and family planning, ensuring the well-being of themselves and their future generations.

Common Reproductive Health Issues and Treatments

With its detailed and delicate balance, the reproductive system is susceptible to various health issues that can affect both males and females. These reproductive health concerns require attention, diagnosis, and appropriate treatment to ensure optimal well-being. Here, we explore some common reproductive health issues and available treatments.

Infertility

Infertility is when a person or couple cannot conceive a child despite regular and unprotected sexual intercourse. It can affect both males and females and may result from various factors such as hormonal imbalances, structural abnormalities, or underlying medical conditions. Infertility treatments include fertility medications, assisted reproductive technologies (ART) like in vitro fertilization (IVF), and surgical interventions.

Sexually Transmitted Infections (STIs)

STIs are infections transmitted through sexual contact and can cause significant reproductive health problems. Common STIs include chlamydia, gonorrhea, syphilis, genital herpes, and human papillomavirus (HPV). Depending on the specific infection, treatment for STIs typically involves antibiotics, antiviral medications, or other targeted therapies.

Polycystic Ovary Syndrome (PCOS)

PCOS is a hormonal disorder affecting females, characterized by enlarged ovaries with small cysts, hormonal imbalances, and irregular menstrual cycles. It can lead to fertility issues, weight gain, acne, and excessive hair growth. Management of PCOS may involve lifestyle changes, hormone therapy, and medications to regulate menstrual cycles and address associated symptoms.

Benign Prostatic Hyperplasia (BPH)

BPH is a non-cancerous prostate gland enlargement that commonly occurs in aging males. It can cause urinary problems, such as frequent urination and weak urine flow. Treatment options for BPH include:

- Medication to alleviate symptoms.
- Minimally invasive procedures.
- Surgery to remove or reduce the size of the prostate gland.

Endometriosis

Endometriosis is a medical condition characterized by the growth of uterine lining tissue outside the uterus, resulting in discomfort, inflammation, and possible challenges with fertility. Various strategies to manage endometriosis encompass pain alleviation, hormonal treatments designed to curtail the proliferation of endometrial tissue, and surgical intervention to excise or minimize the impacted tissue.

Erectile Dysfunction (ED)

ED refers to the inability to achieve or maintain an erection sufficient for sexual intercourse. It can have physical or psychological causes.

Treatments for ED may involve lifestyle changes, counseling, medication, or medical procedures, depending on the underlying cause.

Menstrual Disorders

Menstrual disorders encompass a range of abnormal menstrual patterns, including heavy or prolonged bleeding (menorrhagia), absent or irregular periods (oligomenorrhea), or severe menstrual pain (dysmenorrhea). Treatments may include hormonal therapies, nonsteroidal anti-inflammatory drugs (NSAIDs) for pain management, or surgical interventions in extreme cases.

These are just a few common reproductive health issues that individuals may encounter. It is essential to seek professional medical advice for accurate diagnosis and personalized treatment options based on the specific condition. Early detection, proper management, and access to reproductive healthcare services are crucial in maintaining reproductive health and well-being.

Chapter 7:
The Integumentary System

The integumentary system, our body's exterior armor, is an incredible ensemble of skin, hair, nails, and related glands. This system offers physical protection and plays an instrumental role in temperature balance, sensation, vitamin D synthesis, and immune defense. As we venture into the fascinating realm of the integumentary system, we'll dissect its structure, functions, and astonishing capabilities.

This defensive fortress, the integumentary system, serves as a shield, protecting our bodies from external threats such as pathogens, toxins, and UV radiation. Moreover, it masterfully regulates body temperature through actions like perspiration and the dilation or constriction of blood vessels close to the skin. The skin is further laced with sensory receptors, enabling us to experience sensations like touch, pressure, pain, and temperature variations.

As the body's most extensive organ, the skin is a complex structure composed of three key layers: the epidermis, dermis, and hypodermis (or subcutaneous tissue). The epidermis, the skin's outermost layer, provides a waterproof barrier and is home to melanocytes that produce melanin, giving our skin its unique color. The dermis, located just beneath the epidermis, hosts an array of blood vessels, hair follicles, sweat glands, and sensory receptors. At its deepest level, the hypodermis is a reservoir of fatty tissue providing insulation and energy storage.

The skin's appendages, hair, and nails, constructed of keratinized cells, offer additional layers of protection and functionality. Hair serves multiple roles, providing insulation, acting as a protective barrier, and aiding sensory perception. Nails found on fingertips and toes protect these sensitive areas while assisting in gripping objects.

Complementing the system are the sebaceous glands, producers of sebum, an oily substance that keeps our skin moisturized and bolsters defense against pathogens. Additionally, sweat glands, including the eccrine and apocrine variants, contribute to temperature regulation and waste excretion.

Let's dive deeper into the structure and function of the skin, hair, and nails.

The skin, our protective outer garment, is a complex mosaic of the epidermis, dermis, and hypodermis. Each layer has unique attributes and carries out specific tasks.

The Epidermis

This outer layer of the skin serves as our primary defense against external hazards and UV radiation. It's also a habitat for melanocytes, pigment-producing cells responsible for our skin color, and facilitates skin cell regeneration.

The Dermis

The dermis, nestled beneath the epidermis, provides structural support and nourishment. It's a reservoir of blood vessels, hair follicles, and various glands. It plays a critical role in temperature regulation and sensory perception.

The Hypodermis

Also known as subcutaneous tissue, the hypodermis is the deepest layer of the skin. Composed primarily of fatty tissue, it provides insulation, cushions underlying structures, and stores energy.

Hair, another vital element of the integumentary system, serves various functions, including insulation, protection, and sensory perception, helping us detect environmental changes.

Nails on the fingertips and toes are composed of keratinized cells. They offer protection, assist in handling objects with precision, and, with the help of sensory nerve endings in the nail bed, enable us to detect pressure and touch.

By understanding the structure and functions of the skin, hair, and nails, we gain insight into the remarkable capabilities of the integumentary system. These components work cohesively to protect, regulate, and connect us with our environment, contributing significantly to the well-being of our incredible human bodies.

Common Skin Conditions and Their Treatments

The skin, the outermost protective barrier of our bodies, is susceptible to various conditions and disorders. This section will explore some common skin conditions and the treatments available to address them.

Acne

Acne is a common skin condition characterized by the formation of pimples, blackheads, and whiteheads. It typically occurs during adolescence due to hormonal changes but can affect individuals of any age. Acne treatments include topical creams, cleansers, and oral medications that help reduce inflammation, control oil production, and prevent bacterial growth.

Eczema

Eczema, a chronic inflammatory skin condition, is also known as dermatitis. It results in dry, itchy, and inflamed patches of skin. Moisturizers, corticosteroid creams, and antihistamines are commonly used to manage eczema symptoms and reduce skin inflammation.

Psoriasis

Psoriasis is a chronic autoimmune condition that speeds up the skin cell growth cycle, forming thick, scaly patches on the skin. Topical treatments, phototherapy, systemic medications, and biological therapies alleviate symptoms and manage psoriasis flare-ups.

Dermatitis

Dermatitis is skin inflammation caused by various factors such as irritants, allergens, or genetic predisposition. Treatment involves identifying and avoiding triggers, using corticosteroid creams or ointments, applying moisturizers, and practicing good skincare habits.

Rosacea

Characterized by redness, flushing, and the formation of small blood vessels, rosacea is a chronic inflammatory condition that predominantly affects the face. Treatment options for rosacea include topical medications, oral antibiotics, laser therapy, and lifestyle modifications to manage triggers like sun exposure and certain foods.

Skin Infections

Skin infections, such as bacterial or fungal, can occur due to various factors. Treatment typically involves topical or oral antibiotics or antifungal medications, depending on the specific infection.

Skin Cancer

Skin cancer is a potentially serious condition caused by the abnormal growth of skin cells. Treatment options depend on the type and stage of cancer. Still, they may include surgical removal, radiation therapy, chemotherapy, or targeted therapy.

It is important to note that proper diagnosis and treatment should be sought from qualified healthcare professionals or dermatologists for these and other skin conditions. The treatment approach will depend on the situation, severity, and individual factors.

Understanding common skin conditions and their available treatments enables us to actively care for our skin health, seek timely interventions, and promote overall well-being. Remember, healthy and radiant skin contributes not only to our physical appearance but also to our confidence and self-esteem.

Chapter 8:
The Endocrine System

Embarking on a fascinating exploration of our body's internal communication network, we delve into the realm of the endocrine system. This complex maze of glands, responsible for producing and disseminating hormones, masterfully choreographs various vital bodily functions.

The endocrine system deftly ensures homeostasis, guiding numerous physiological processes. It deftly modulates our growth, metabolism, reproduction, mood, and overall function. With this robust network, our bodies would gain the direction to maintain optimal health.

Let's discover more about the endocrine system's maestros - the glands. Scattered throughout our body, they form a stellar cast, including the pituitary, thyroid, adrenal glands, pancreas, parathyroid, pineal gland, ovaries (in females), and testes (in males). Each gland meticulously produces specific hormones, their roles defined by their unique effects on target organs and tissues.

Hormones, the endocrine system's secret messages, traverse through our bloodstream to their destination cells or organs. They bind to specific receptors on reaching, setting off a cascade of physiological responses. These potent substances influence many functions, including metabolism, growth, development, reproduction, stress responses, blood sugar regulation, and electrolyte balance.

The hypothalamus, a humble region in the brain, creates a crucial bridge between the nervous and endocrine systems. It directs the pituitary gland, often celebrated as the 'master gland.' The pituitary, in turn, regulates growth and sexual development and releases a gamut of hormones controlling various body processes.

The thyroid and parathyroid glands regulate our metabolism, energy production, body temperature, and blood calcium levels. Meanwhile, our adrenal glands, perched atop our kidneys, manage stress response, metabolism, salt and water balance, and blood pressure regulation.

The pancreas plays dual roles as an endocrine and exocrine gland, ensuring blood sugar levels and energy storage balance. Any malfunction in this crucial gland can result in diabetes mellitus, impaired insulin production, or insulin resistance.

With an enriched understanding of the endocrine system, we'll truly appreciate its pivotal role in choreographing our growth, metabolism, reproduction, and the delicate dance of our hormonal balance. So, hop on board for this enlightening expedition as we navigate the winding trails and extraordinary influences of hormones on our bodies and minds.

Regulation of Bodily Functions through Hormone Signaling

Hormones are vital in regulating diverse physiological functions and promoting overall homeostasis. This section will explore how hormone signaling regulates different physiological processes throughout the body.

Hormone Signaling

Hormones function as chemical messengers released by endocrine glands, traveling through the bloodstream. Once in the bloodstream, they reach target cells or tissues, binding to specific receptors. This binding triggers a cascade of biochemical reactions within the cells, ultimately leading to specific physiological responses.

Feedback Mechanisms

The regulation of hormone secretion is tightly controlled by feedback mechanisms to maintain balance within the body. Two primary types of feedback mechanisms exist:

- **Negative Feedback:** When hormone levels reach a certain threshold, negative feedback mechanisms inhibit further hormone secretion. This helps prevent excessive hormone production and maintains stability within the body.
- **Positive Feedback:** Positive feedback mechanisms sometimes amplify a physiological response. This occurs when the initial stimulus promotes increased hormone secretion, further enhancing the reaction.

Hormonal Regulation of Bodily Functions

The endocrine system, through its intricate network of hormones, influences a wide range of bodily functions, including:

- **Metabolism:** Hormones such as insulin, glucagon, thyroid hormones, and cortisol regulate metabolism by controlling energy production, glucose utilization, and lipid metabolism.

- **Growth and Development:** Growth hormones, thyroid hormones, and sex hormones play critical roles in regulating growth, development, and sexual maturation.

- **Reproduction:** Hormones like estrogen, progesterone, testosterone, follicle-stimulating hormone (FSH), and luteinizing hormone (LH) regulate reproductive processes, including fertility, menstruation, and pregnancy.

- **Stress Response:** Hormones like cortisol and adrenaline (epinephrine) are involved in the body's response to stress, enabling it to cope with challenging situations.

- **Calcium Balance:** Parathyroid hormone (PTH) and calcitonin regulate calcium levels in the blood, ensuring proper bone health and nerve function.

- **Water and Electrolyte Balance:** Hormones such as aldosterone and antidiuretic hormone (ADH) help maintain the body's balance of water and electrolytes by influencing kidney function.

Imbalances and Disorders of the Endocrine System

While the endocrine system maintains balance and proper functioning within the body, imbalances and disorders can disrupt this delicate equilibrium. This section will explore common imbalances and disorders of the endocrine system and their effects on overall health.

Hormone Imbalances

Hormone imbalances can occur when there is either an excess or deficiency of specific hormones in the body. These imbalances can arise from various factors, including genetic predispositions, lifestyle choices, environmental factors, or underlying medical conditions. Some common hormone imbalances include:

- **Hyperthyroidism:** An overactive thyroid gland leading to excessive production of thyroid hormones, resulting in symptoms such as weight loss, rapid heartbeat, and anxiety.

- **Hypothyroidism:** An underactive thyroid gland leading to insufficient production of thyroid hormones, causing symptoms such as fatigue, weight gain, and depression.

- **Diabetes:** A disorder characterized by high blood sugar levels due to either insufficient insulin production (Type 1 diabetes) or impaired insulin function (Type 2 diabetes).

- **Adrenal Insufficiency:** Occurs when the adrenal glands do not produce enough cortisol and aldosterone, leading to fatigue, low blood pressure, and electrolyte imbalances.

- **Pituitary Disorders:** Abnormalities in the pituitary gland can cause disruptions in various hormone production, leading to conditions like growth hormone deficiency or pituitary tumors.

- **Polycystic Ovary Syndrome (PCOS):** A hormonal disorder in females that can cause irregular menstrual cycles, excessive hair growth, and fertility issues.

Endocrine Disorders

Apart from hormone imbalances, specific disorders affect the endocrine system as a whole. These disorders can impact the function of multiple glands or disrupt hormone production and regulation. Some examples include:

- **Cushing's Syndrome:** A condition characterized by excessive cortisol levels in the body, leading to weight gain, muscle weakness, and mood changes.

- **Addison's Disease:** Manifests when the adrenal glands fail to generate sufficient amounts of cortisol and aldosterone, leading to symptoms such as fatigue, low blood pressure, and electrolyte imbalance.

- **Hypopituitarism:** A condition where the pituitary gland fails to produce one or more hormones, causing a range of symptoms depending on the affected hormones.

Impact on Health and Treatment

Imbalances and disorders of the endocrine system can significantly affect overall health and well-being. They can disrupt metabolism, growth and development, reproductive functions, and other physiological processes. Proper diagnosis and treatment are crucial in managing these conditions. Depending on the disorder's severity, treatment options may include hormone replacement therapy, lifestyle modifications, medication, or surgery.

Understanding the imbalances and disorders within the endocrine system is vital in recognizing symptoms, seeking appropriate medical care, and maintaining optimal health.

Chapter 9:
The Immune System

Our bodies are home to a fascinating defense network - the immune system - consisting of organs, cells, and molecules that collaborate harmoniously to protect us from harmful invaders like bacteria, viruses, and parasites. Prepare yourself for a captivating exploration of this labyrinth, where we decode its critical functions in safeguarding our health and maintaining our well-being at its zenith.

Think of the immune system as an expert military defense unit with two primary strategies: the innate and the adaptive immune system. The inherent division serves as our body's rapid, broad-spectrum defense force. In contrast, the adaptive division develops targeted, long-lasting immunity against specific pathogens.

White blood cells - the soldiers of our immune army - come in various types, such as neutrophils, lymphocytes, and macrophages, each playing their part in detecting and eliminating pathogens. Lymphocytes, including T and B cells, are the linchpins of our adaptive immune responses.

But what's an army without a fortress? The immune system's stronghold is an assortment of lymphoid organs like the thymus, spleen, and lymph nodes, where immune cells mature and prepare for action. Other important components include antibodies - specialized proteins that mark specific pathogens for destruction - and cytokines, the communication channels that regulate immune responses.

The immune system responds to invading pathogens like a well-coordinated military operation. It starts with pathogen recognition and immune cell activation, progresses to inflammation, and culminates in a specific T and B-cell-mediated response. This response doesn't just fight the present invasion; it also prepares us for future encounters with the same enemy – a remarkable quality we leverage in vaccinations to provide long-term immunity.

However, like all things complex, the immune system is full of challenges. Autoimmune diseases occur when it mistakenly attacks our tissues, while immunodeficiency disorders weaken the immune response, making us more susceptible to infections. We can support and maintain a robust immune system through mindful practices like proper nutrition, adequate sleep, stress management, and healthy lifestyle choices.

In this chapter, we'll dive deeper into the complex labyrinth of the immune system. We'll explore the key mechanisms, components, and astounding ways this biological marvel keeps us safe from diseases. Additionally, we'll discuss the exciting domain of immunology to reveal breakthroughs in research and a glimpse at the future of immune-based therapies and treatments.

Role of the Immune System in Defending Against Pathogens

Innate Immunity

The first line of defense is the innate immune response, which provides immediate, nonspecific protection against various pathogens. Key components of innate immunity include:

- **Physical Barriers:** The skin and mucous membranes act as physical barriers, preventing pathogens from entering the body. They secrete substances that inhibit the growth of microorganisms.

- **Phagocytes:** Phagocytes, including neutrophils and macrophages, engulf and destroy pathogens through phagocytosis.

- **Natural Killer Cells:** Natural killer cells detect and destroy virus-infected cells and cancerous cells, helping to prevent the spread of infection.

- **Inflammatory Response:** Inflammation is a localized reaction triggered by tissue injury or infection, characterized by heightened blood flow, the recruitment of immune cells, and the release of inflammatory mediators. These processes collectively aim to eliminate pathogens and facilitate the healing process.

Adaptive Immunity

The adaptive immune response is a highly specialized defense mechanism that develops over time and provides long-term immunity against specific pathogens. It involves two key components:

- **B Cells and Antibodies:** B cells produce antibodies that recognize and bind to specific antigens in pathogens. These antibodies neutralize the pathogens, mark them for destruction, and activate other immune cells to eliminate them.

- **T Cells:** T cells assume a vital role in cell-mediated immunity. They can directly recognize and eliminate infected cells, stimulate B cells to produce antibodies or regulate the immune response.

Immune Memory

One of the most remarkable features of the immune system is its ability to develop immunological memory. Upon encountering a pathogen for the first time, the immune system mounts a primary immune response. Subsequent encounters with the same pathogen lead to a faster and stronger secondary immune response due to the presence of memory B and T cells. This memory response provides long-lasting protection against reinfection.

Immunization

Immunization, or vaccination, is a powerful tool that harnesses the immune system's ability to develop memory responses. Vaccines contain harmless antigens that stimulate an immune response, allowing the body to recognize and remember the pathogen. This way, if the individual encounters the pathogen in the future, their immune system can mount a rapid and effective response, preventing disease development.

Autoimmune Diseases and Immune System Disorders

While the immune system's primary function is to shield the body from external threats, there are instances when it may malfunction and erroneously target its cells and tissues. These conditions are known as autoimmune diseases and immune system disorders, and they can profoundly impact an individual's health and well-being. This section will explore some common autoimmune diseases and immune system disorders.

Autoimmune Diseases

Autoimmune diseases occur when the immune system mistakenly identifies normal, healthy cells as foreign and launches an immune response against them. Some well-known autoimmune disorders include:

- **Rheumatoid Arthritis:** This chronic inflammatory disease primarily affects the joints, causing pain, swelling, and stiffness.
- **Systemic Lupus Erythematosus:** Lupus is a systemic autoimmune disease that can affect multiple organs and tissues, leading to symptoms such as fatigue, joint pain, skin rashes, and kidney problems.

- **Multiple Sclerosis:** In this autoimmune condition, the immune system launches an assault on the protective coating of nerve fibers in the central nervous system, causing disruptions in the communication between the brain and the rest of the body.

- **Type 1 Diabetes:** Type 1 diabetes arises when the immune system erroneously targets and destroys the insulin-producing cells within the pancreas, leading to elevated blood sugar levels.

Allergies

Allergies are hypersensitivity reactions triggered by exposure to harmless substances, such as pollen, dust mites, or certain foods. In individuals with allergies, the immune system overreacts to these substances, leading to symptoms like sneezing, itching, hives, and in severe cases, anaphylaxis.

Immunodeficiency Disorders

Immunodeficiency disorders occur when the immune system is weakened or compromised, leaving the body susceptible to infections. Some examples of immunodeficiency disorders include:

- **Primary Immunodeficiency Disorders:** These are genetic conditions that result in a weakened immune system from birth, making individuals more susceptible to infections.

- **Acquired Immunodeficiency Syndrome (AIDS):** AIDS is the consequence of infection by the human immunodeficiency virus (HIV), which targets and eradicates the CD4 cells of the immune system, leaving the body susceptible to opportunistic infections.

Hypersensitivity Reactions

Hypersensitivity reactions are exaggerated immune responses to specific substances. The immune system reacts as if the substance is harmful, triggering an immune response that can cause symptoms ranging from mild to severe. Examples include allergic rhinitis (hay fever), asthma, and drug allergies.

Autoinflammatory Disorders

Autoinflammatory disorders are characterized by recurrent episodes of inflammation in the absence of an autoimmune or infectious cause. These conditions result from a dysfunction in the innate immune system, leading to uncontrolled inflammation. Examples include familial Mediterranean fever and periodic fever syndromes.

Understanding autoimmune diseases and immune system disorders is crucial for early detection, diagnosis, and management. The continuous progress in research and treatment options provides optimism for individuals grappling with these conditions, striving to enhance their quality of life and overall well-being.

Chapter 10:
The Urinary System

The urinary system, also known as the renal system, plays a vital role in maintaining the body's internal balance by regulating fluid levels, electrolytes, and waste elimination. Comprised of several organs working together, the urinary system ensures the proper functioning of the body's filtration and excretion processes.

Overview of the Urinary System

Our bodies are complex and elaborate systems with myriad subsystems collaborating seamlessly. Among these, the urinary system stands out as an impressive biological marvel. This essential network, composed of the kidneys, ureters, bladder, and urethra, tirelessly purges waste from our bodies and regulates our water balance, ensuring optimal health and hydration.

At the heart of this system lie the kidneys, our sophisticated filtration and reabsorption centers. The kidneys, nestled within the upper abdominal cavity, are bean-shaped powerhouses that contain millions of tiny nephrons, which are the building blocks of the filtration process. Beyond this pivotal task, the kidneys balance fluid and electrolyte levels, stabilize blood pressure, and produce hormones that manage red blood cell production and calcium metabolism.

The fascinating journey of urine formation commences in the nephrons. As blood courses into the kidneys through the renal arteries, the nephrons meticulously sieve out waste, excess water, and electrolytes. The resulting primary urine still contains nutrients and water, diligently reabsorbed into the bloodstream, maintaining the body's delicate equilibrium. The remaining purified urine then journeys to the bladder, a reservoir awaiting expulsion, marking the end of a sophisticated cycle of filtration and regulation.

Now, let's delve deeper into each component of the urinary system, starting with the kidneys. Located on either side of the spine in the upper abdomen, they are the unsung heroes regulating our body's water balance, electrolytes, and acids while helping control blood pressure and stimulating red blood cell production.

Continuing our exploration, we encounter the ureters, which function as slender ducts that act as conduits, transporting urine from the kidneys to the bladder. To ensure a one-way flow, the smooth muscle walls of the ureters undergo rhythmic contractions.

The bladder, a muscular sac situated behind the pubic bone in the lower abdomen, provides temporary storage for urine. This unique structure, lined with transitional epithelium, expands and contracts to accommodate varying volumes of urine, preventing leakage.

Lastly, the urethra is the final pathway that conveys the urine from the bladder out of the body during urination. In males, this tube also serves as a conduit for semen during ejaculation. Its length varies between the sexes, being typically longer in males.

Collectively, the kidneys, ureters, bladder, and urethra form a harmonious ensemble, ensuring the urinary system's proper functioning. The kidneys filter waste and excess fluids from the blood, yielding urine that the ureters channel to the bladder. This temporary storage vessel holds the urine until it's voluntarily excreted during urination.

The urinary system plays a crucial role in waste elimination and maintaining the body's fluid, electrolyte, and acid-base balance. The kidneys keep electrolyte concentrations, including sodium, potassium, and calcium, within a healthy range, essential for optimal physiological functioning.

Post-filtration, the renal tubules reclaim vital substances like water, glucose, and electrolytes, transporting them back into the bloodstream through surrounding capillaries. Simultaneously, additional waste is secreted into the tubules. This selective process preserves essential substances while maintaining fluid balance.

Further aiding balance, the kidneys regulate urine concentration and dilution via hormonal signals that adjust renal tubule permeability. When the body needs to conserve water, the tubules reabsorb more, resulting in concentrated urine. Conversely, when water needs to be expelled, the tubules reabsorb less, diluting the urine.

This complex yet efficient filtration and waste elimination process is fundamental in preserving the body's internal balance and expelling harmful substances, ensuring our body functions properly. In the upcoming sections, we'll explore common disorders and diseases affecting the urinary system, along with the available diagnostic techniques and treatment options.

Common Urinary System Disorders and Their Treatments

The urinary system can be susceptible to various disorders and conditions affecting normal functioning. This section will explore some common urinary system disorders and the available treatments for each.

Urinary Tract Infections (UTIs)

UTIs occur when bacteria enter the urinary tract and multiply, leading to infection. The most common UTIs affect the bladder (cystitis) or the urethra (urethritis). Possible symptoms include increased urination frequency, a burning sensation while urinating, urine that appears cloudy or bloody, and pelvic pain. UTIs are typically treated with antibiotics to eliminate the bacterial infection. Drinking plenty of water and practicing good hygiene can help prevent UTIs.

Kidney Stones

Kidney stones are hard mineral and salt deposits that form in the kidneys. As they traverse the urinary tract, they can induce intense pain. Treatment options for kidney stones depend on their size and location. Small stones may pass naturally with increased fluid intake and pain management. Larger stones may require medical intervention, such as extracorporeal shock wave lithotripsy (ESWL), ureteroscopy, or surgical removal.

Urinary Incontinence

Urinary incontinence is characterized by the unintentional release of urine. It can result from weakened pelvic floor muscles, nerve damage, or underlying medical conditions. Treatment options for urinary incontinence include pelvic floor exercises (Kegel exercises), bladder training techniques, medication, and in severe cases, surgery. Making lifestyle adjustments, such as steering clear of bladder irritants and maintaining a healthy weight, can also aid in managing the condition.

Urinary Retention

Urinary retention occurs when the bladder cannot empty completely, leading to a persistent feeling of incomplete bladder emptying. It can be caused by various factors, including bladder muscle dysfunction, obstruction, or nerve damage. Treatment options for urinary retention depend on the underlying cause and may include medication to relax the bladder muscles, catheterization to empty the bladder, or surgical procedures to address structural issues.

Bladder Cancer

Bladder cancer refers to the abnormal growth of cells in the bladder lining. Symptoms may include blood in the urine, frequent urination, and pain during urination. Treatment options for bladder cancer depend on the stage and grade of the tumor. They may involve surgery, radiation therapy, chemotherapy, or immunotherapy. Early detection and prompt treatment are crucial for better outcomes.

Urinary System Infections and Inflammation

Apart from UTIs, other infections and inflammatory conditions can affect the urinary system, such as interstitial cystitis, prostatitis, or pyelonephritis. Treatment approaches may include antibiotics, anti-inflammatory medications, pain management, and lifestyle modifications to alleviate symptoms and manage the underlying causes.

It's important to note that each urinary system disorder is unique, and treatment plans should be tailored to the individual's specific condition and medical history. Suppose you experience any symptoms or concerns related to the urinary system. In that case, it is advisable to consult a healthcare professional for an accurate diagnosis and appropriate treatment.

Chapter 11:
Mind-Blowing Facts and Discoveries

The human body continues to be a subject of awe and fascination, and scientific research has led to numerous remarkable discoveries and advancements. This chapter will delve into some impressive facts and breakthroughs that have revolutionized our understanding of the human body.

Genomic Sequencing

Completing the Human Genome Project in 2003 was a monumental achievement that provided a comprehensive map of the human genetic code. This breakthrough has enabled scientists to study the relationships between genes, diseases, and inherited traits, opening doors to personalized medicine and targeted therapies.

Neuroplasticity

The discovery of neuroplasticity shattered the long-held belief that the brain is a fixed and unchanging organ. Research has shown that the brain can adapt and rewire itself, forming new neural connections and pathways throughout life. This finding has profound implications for brain rehabilitation, learning, and understanding various mental health conditions.

Gut-Brain Connection

Emerging research has revealed the relationship between the gut and the brain, known as the gut-brain axis. The gut houses trillions of microorganisms, collectively known as the gut microbiota, which play a crucial role in digestion and metabolism. Moreover, these microbes communicate with the brain through complex pathways, influencing mood, cognition, and even behavior. Understanding this connection opens up new avenues for treating mental health disorders and gastrointestinal conditions.

Regenerative Medicine

Advancements in regenerative medicine have shown promise in repairing and replacing damaged tissues and organs. Stem cell research, tissue engineering, and 3D printing technologies have paved the way for regenerating tissues such as skin, cartilage, and even organs like the liver and heart. These breakthroughs offer hope for patients with debilitating injuries or organ failure.

CRISPR-Cas9 Gene Editing

The development of CRISPR-Cas9 gene editing technology has revolutionized the field of genetics. This powerful tool allows scientists to precisely edit genes, opening up possibilities for treating genetic disorders, modifying traits, and potentially curing diseases. CRISPR can potentially reshape the future of medicine and has garnered significant attention in the scientific community.

Microbiome Research

The human microbiome, consisting of the vast community of microorganisms inhabiting our bodies, has emerged as a fascinating study area. Research has shown that these microbes influence our health, metabolism, immune system, and mental well-being. Understanding the complexity of the microbiome has the potential to transform healthcare practices and lead to innovative treatments.

Artificial Intelligence in Healthcare

Artificial intelligence (AI) has made significant strides in healthcare, aiding diagnosis, treatment planning, and data analysis. Machine learning algorithms and AI systems can process vast amounts of medical data, leading to more accurate diagnoses, personalized treatment recommendations, and improved patient outcomes.

These remarkable discoveries and advancements are just a glimpse of the ongoing research and exploration of the human body. As scientists continue to uncover the mysteries of our spectacular biology, the potential for new breakthroughs and transformative discoveries is limitless.

Surprising Facts and Mysteries about the Human Body

The human body is a remarkable creation full of astonishing features and mysteries yet to be fully understood. This chapter will explore some surprising facts and delve into the enigmatic aspects that continue to captivate scientists and researchers worldwide.

Superhuman Strength

Under extreme circumstances, individuals have displayed incredible feats of strength. Adrenaline, the "fight or flight" hormone, can temporarily boost strength and endurance, allowing individuals to lift heavy objects or perform remarkable physical tasks. The extent of this hidden strength potential is still a subject of fascination and exploration.

Phantom Limb Sensation

People who have undergone amputation sometimes experience the sensation that their missing limb is still present. This phenomenon, known as phantom limb sensation, is believed to be caused by the brain's inability to fully adjust to the absence of the limb. The brain continues receiving signals from the nerves previously connected to the amputated limb, creating a vivid and perplexing experience.

Unique Fingerprints

No two individuals have the same fingerprints. Even identical twins share the same DNA and have distinct fingerprint patterns. The ridges and loops on our fingertips are formed during fetal development and remain unchanged throughout our lives. The exact reason behind the uniqueness of fingerprints remains a fascinating mystery.

Sleep Paralysis

Sleep paralysis is when a person is temporarily unable to move or speak while falling asleep or waking up. It occurs due to a disruption in the transition between sleep stages, leaving individuals briefly conscious but unable to control their muscles. This intriguing phenomenon has sparked various cultural interpretations and has been linked to paranormal experiences and hallucinations.

Handedness

Approximately 90% of people are right-handed, while 10% are left-handed or ambidextrous. The reasons behind handedness still need to be fully understood. Still, genetic, environmental, and cultural factors are believed to influence it. The dominance of one hand over the other adds an intriguing dimension to the complexity of human behavior.

Biological Clocks

The human body operates on various internal biological clocks that regulate essential functions such as sleep-wake cycles, hormone production, and body temperature. The body's internal timekeepers, called circadian rhythms or biological clocks, are impacted by external elements like light and hold significant importance in sustaining overall health and well-being. The intricacies of these internal timekeepers continue to be an area of active research.

Human Longevity

While the human lifespan has limitations, researchers are continually exploring the factors that contribute to exceptional longevity. Certain regions of the world, known as Blue Zones, have a higher concentration of individuals living beyond 100 years. Genetics, lifestyle choices, and environmental factors are believed to play a role in these remarkable cases, providing insights into the possibilities of extending human life.

Cutting-Edge Technologies and Techniques Used in Studying the Human Body

The exploration of the human body has been greatly enhanced by advancements in technology and innovative research techniques. This section will delve into the cutting-edge tools and methodologies that have revolutionized our understanding of the human body.

Magnetic Resonance Imaging (MRI)

MRI technology utilizes powerful magnetic fields and radio waves to generate detailed images of the body's internal structures. It provides a non-invasive and highly accurate way to visualize organs, tissues, and even the brain. MRI has become a cornerstone of medical diagnostics and research, enabling us to examine the human body in unprecedented detail.

Genomic Sequencing

The Human Genome Project marked a significant milestone in our understanding of human genetics. Today, genomic sequencing techniques have become more advanced and affordable, allowing scientists to decode an individual's entire genetic blueprint. This breakthrough technology has opened up new avenues for personalized medicine, disease research, and the exploration of our genetic heritage.

Electron Microscopy

Electron microscopy enables scientists to visualize structures at an incredibly high resolution. Researchers can examine tissues, cells, and even individual molecules with extraordinary precision using a beam of accelerated electrons. This technique has unveiled details of cellular structures and processes, providing invaluable insights into the complexity of the human body.

Optogenetics

Optogenetics combines optics and genetics to manipulate and observe the activity of specific cells in living organisms. By introducing light-sensitive proteins into targeted cells, researchers can control their function with precise timing. This powerful technique has shed light on the complex workings of the brain. It holds promise for understanding neurological disorders and developing new therapeutic approaches.

Functional Magnetic Resonance Imaging (fMRI)

fMRI measures changes in blood flow within the brain to map brain activity and identify functional areas associated with specific tasks or stimuli. It provides insights into how different brain regions are interconnected and work together. fMRI has been instrumental in discovering the mysteries of cognition, emotion, and neurological disorders.

Single-Cell Analysis

Advancements in single-cell analysis techniques have enabled scientists to study individual cells with unprecedented detail. By analyzing a single cell's genetic, molecular, and functional characteristics, researchers can uncover cellular diversity, understand cell development, and study disease mechanisms at a cellular level. This approach has the potential to revolutionize our understanding of complex biological systems.

These cutting-edge technologies and techniques are propelling human body research into new frontiers. They empower scientists to unravel the complexities of our anatomy, physiology, and genetic makeup, leading to breakthrough discoveries and transformative advancements in medicine and healthcare.

Conclusion

Throughout this captivating exploration of the human body in "The Abyss Inside: Mind-Blowing Facts and Discoveries About Your Extraordinary Human Body," we have embarked on an awe-inspiring journey of discovery. From the compounded systems that sustain our existence to the wondrous mechanisms that enable our every thought, movement, and sensation, we have explored the depths of our own miraculous biology. As we conclude this enlightening journey, let us recap the key points and insights we have gained.

Anatomy

We have unraveled the marvels of human anatomy, comprehending the structure, organization, and interconnectedness of our bones, muscles, nerves, and organs. The human body is a detailed masterpiece, with each component crucial in maintaining our well-being.

Circulatory System

We have marveled at the circulatory system, the vital transportation network that delivers oxygen, nutrients, and essential substances to every cell and removes waste products. We have explored the pulmonary and systemic circuits, gaining an understanding of how blood flows and its crucial role in maintaining equilibrium.

Digestive System

We have embarked on a journey through the digestive system, witnessing the transformation of food into nourishment. We have discovered the significance of each organ, from the mouth to the intestines, and the remarkable digestion and nutrient absorption process.

Nervous System

We have unraveled the mysteries of the nervous system, the complex network that enables communication and coordination throughout our body. We have explored the structure and function of neurons and glial cells, delving into the central and peripheral nervous systems and their roles in sensory perception and motor control.

Reproductive System

We have explored the wonders of the reproductive system, understanding the unique processes of male and female anatomy, puberty, menstruation, and fertility. We have contemplated the miraculous creation of life and the balance of hormones that orchestrate reproduction.

Integumentary System

We have appreciated the remarkable functions of the integumentary system, including the skin, hair, and nails, which provide protection, regulation, and sensory perception. We have learned about the skin's role as a sensory interface with the world and the significance of its health and care.

Endocrine System

We have delved into the endocrine system, unraveling the vital role of hormones in regulating bodily functions. We have explored the major endocrine glands and their profound impact on growth, metabolism, reproduction, and overall well-being.

Immune System

We have marveled at the immune system, our impressive defense against pathogens and foreign invaders. We have learned about immune response mechanisms and the significance of white blood cells and lymph nodes in maintaining our health.

Urinary System

We have discovered the remarkable role of the urinary system in maintaining fluid balance, filtering waste products, and regulating blood pressure. We have explored the anatomy and functions of the kidneys, bladder, ureters, and urethra.

Fascinating Discoveries

We have been enthralled by the fascinating discoveries and advancements in human body research. Throughout our journey, we have encountered astonishing revelations, enigmatic phenomena, and state-of-the-art technologies that consistently challenge the limits of our comprehension.

"The Abyss Inside: Mind-Blowing Facts and Discoveries About Your Extraordinary Human Body" has unveiled our mysteries, awakening a deep sense of wonder and appreciation for the incredible vessel that houses our consciousness. May this knowledge inspire us to nurture and care for our bodies, celebrate their resilience and adaptability, and embark on a lifelong journey of exploration and self-discovery.

Encouragement to Continue Exploring and Appreciating the Wonders of the Human Body

As we reach the end of our journey through "The Abyss Inside: Mind-Blowing Facts and Discoveries About Your Extraordinary Human Body," let us not view this as the culmination of our exploration but rather as a stepping stone towards a lifelong appreciation and fascination with the wonders of the human body.

The knowledge we have acquired about our anatomy, physiology, and interwoven systems is just the beginning. It is an invitation to dive deeper, to continue seeking knowledge and understanding about the complexities that make us who we are.

Every day, remarkable discoveries and advancements are being made in human biology. New insights emerge, unraveling further mysteries and expanding our understanding of the intricacies of our existence. Embrace this spirit of curiosity and embark on a personal journey of exploration.

Take the time to listen to your body, to be attuned to its signals and needs. Recognize the remarkable synchronicity between your thoughts, emotions, and physical sensations. Cherish the awe-inspiring moments when you witness the body's resilience, adaptability, and capacity for healing.

Engage in conversations with experts, researchers, and fellow enthusiasts who share your passion for the human body. Attend lectures, read books, and explore online resources to stay abreast of the latest discoveries and developments. Seek out opportunities to participate in studies or contribute to ongoing research projects.

Engaging in physical activities, such as exercise, dance, or yoga, allows us to connect with our bodies profoundly. By nurturing our physical well-being, we foster a deeper understanding and appreciation for the marvels that lie within.

Moreover, let us recognize the interconnectedness of our bodies with the natural world. Appreciate the role of nutrition, rest, and environmental factors in maintaining optimal health. Embrace a holistic approach that recognizes the synergy between our bodies, minds, and the world we inhabit.

As we conclude our exploration of the human body, remember that the journey does not end here. It is an ongoing quest, an opportunity to deepen our understanding and a lifelong celebration of the magnificent vessel that carries us through life.

May the knowledge and insights gained from "The Abyss Inside: Mind-Blowing Facts and Discoveries About Your Extraordinary Human Body" inspire you to continue exploring, appreciating, and nurturing the wonders of your own body. Embrace the gift of self-discovery and let it fuel your curiosity, awe, and reverence for the magnificent creation that is the human body.

We've reached the end of our journey through "The Infinite Abyss Series: Ignite Your Passion for Learning," and I hope that your curiosity has been ignited and your passion for learning about our world, the depths of the ocean, the human body, and the vast cosmos has been stimulated. Our voyage together might have concluded, but the beauty of knowledge is that there's always more to discover, explore, and learn.

Your feedback is precious and can help other explorers set the course on their learning journey. Please share your thoughts about this voyage by leaving a review on Amazon. Did you uncover fascinating new facts? Was your curiosity piqued? How has your perspective changed? Your review can guide future readers, helping them decide if this series is the right launchpad for their journey into the infinite abyss of knowledge.

To leave a review, visit the book's page on Amazon, scroll down to the customer reviews, and click 'Write a customer review.' Your insights and experiences can make all the difference for someone else on the cusp of their journey of discovery.

Resources

AAFP. (n.d.). American Academy of Family Physicians. http://www.aafp.org/

Access anytime anywhere | Cleveland Clinic. (n.d.). Cleveland Clinic. https://my.clevelandclinic.org/

American Heart Association | To be a relentless force for a world of longer, healthier lives. (n.d.). www.heart.org. http://www.heart.org/

CDC works 24/7. (2023, July 19). Centers for Disease Control and Prevention. http://www.cdc.gov/

Fish, T. (2021, October 27). 12 Mind-Blowing Facts about your Body. *Newsweek.* https://www.newsweek.com/mind-blowing-facts-about-your-body-human-1638872

Home. (2023, July 18). http://www.who.int/

https://website-designer-2149.business.site/. (n.d.). *WEIGHT-CUTTING, PEDs, AND TRT - The Ultimate Guide to Preventing and Treating MMA Injuries: Featuring advice from UFC Hall of Famers Randy Couture, Ken Shamrock, Bas Rutten, Pat Miletich, Dan Severn and more! (2016).* Publicism - Non-fiction Documentary Books. https://publicism.info/sports/mma/9.html

Information and Resources about for Cancer: Breast, Colon, Lung, Prostate, Skin. (n.d.). American Cancer Society. http://www.cancer.org/

Interesting facts about the human body | SelectHealth. (2020, February 15). SelectHealth.org.

https://selecthealth.org/blog/2020/02/15-interesting-facts-about-the-human-body

MedlinePlus. (n.d.). *MedlinePlus - Health Information from the National Library of Medicine*. https://medlineplus.gov/

National Institutes of Health (NIH). (n.d.). National Institutes of Health (NIH). http://www.nih.gov/

Neurological disorders. (2023, May 3). Johns Hopkins Medicine. https://www.hopkinsmedicine.org/health/conditions-and-diseases/neurological-disorders

Ng, A. (2023, June 29). *15 Facts About The Human Body! - National Geographic Kids*. National Geographic Kids. https://www.natgeokids.com/uk/discover/science/general-science/15-facts-about-the-human-body/

No. 1 Hospital in the Nation – Mayo Clinic. (n.d.). Mayo Clinic. http://www.mayoclinic.org/

Pjallens. (2023, January 19). Paul Presents The Berkshire MS Therapy Centre with Kim Williams | West Berkshire Villagers. *West Berkshire Villagers*. https://westberksvillagers.com/paul-presents-the-berkshire-ms-therapy-centre-with-kim-williams/

Top 10 diseases of the nervous system. (2023, March 15). https://db4phone.com/en/post/top-10-diseases-of-the-nervous-system.html

Tpt, T. P. (n.d.). *Roshiita | Nephrology and Urology I Book now Kidney-and-urinary-tract-diseases - in Algeria*. https://roshiita.com/en/Algeria/Blog/Kidney-and-urinary-tract-diseases

Vanstone, E. (2022). Human body Facts. *Science Experiments for Kids*. https://www.science-sparks.com/human-body-facts/

WebMD - Better information. Better health. (2023, July 19). WebMD. http://www.webmd.com/

What Are The Signs of Central Precocious Puberty (CPP)? (n.d.). https://www.pubertytoosoon.com/about-cpp/what-are-the-signs-of-central-precocious-puberty

Wikipedia contributors. (2023d). Human body. *Wikipedia.* https://en.wikipedia.org/wiki/Human_body